The Bristol Coalfield

John Cornwell

Published by

Landmark Publishing Ltd
Ashbourne Hall, Cokayne Ave, Ashbourne, Derbyshire DE6 1EJ England
Tel: (01335) 347349 Fax: (01335) 347303
e-mail: landmark@clara.net
web site: www.landmarkpublishing.co.uk

ISBN 1 84306 094 9

© **John Cornwell 2003**

The rights of the author to this work has
been asserted by him in accordance with the Copyright,
Design and Patents Act, 1993.

All rights reserved. No part of this publication may be reproduced, stored in a retrieval system or transmitted in any form or by any means, electronic, mechanical, photocopying, recording or otherwise without the prior permission of Landmark Publishing Ltd. For the avoidance of doubt, this includes reproduction of any image in this book on an internet site.

British Library Cataloguing in Publication Data: a catalogue
record for this book is available from the British Library.

Print: Bath Press, Bath
Design: Mark Titterton
Cover: James Allsopp

Front cover: Speedwell Colliery around 1899 with a wooden headframe.

Title page: Easton Colliery taken around 1900.

Back cover: The Hanham Colliery rescue team.

LANDMARK COLLECTOR'S LIBRARY

THE BRISTOL COALFIELD

John Cornwell

Landmark Publishing

Contents

Foreword	5
1. The Coalfield	**7**
Mining in the 17th, 18th and early 19th centuries	7
The water levels of the Kingswood and Fishponds area	21
Mine Lighting	21
The Newcomen Engine	23
An engine house for a Newcomen engine on south side of Argyle Road, Chester Park	24
Known Newcomen engines in Kingswood, Bristol area	25
2. The Western Pits	**32**
Easton Colliery	32
Whitehall Colliery	35
Pilemarsh Collieries	37
The Great Western Colliery	38
Mines in Crews Hole	40
3. The Kingswood Pits	**44**
Two Mile Hill Collieries	44
Lodge Hill and Hillfields	46
Fishponds and Staple Hill	49
Kingswood and Hanham	56
Hanham Colliery	51
Mines in the Cock Road area	55
The Kingswood Collieries: Speedwell and Deep Pit	56
Rotherham Colliery	67
Soundwell Colliery	67
The New Cheltenham Pit	70
Potters Wood and Jays Pits	70
California Collieries	74
Hole Lane and Bull Hall Collieries	76
Cowhorn Hill, Brook and Buff Pits	78
Goldney Pit, Cadbury Heath	78
The Grimsbury Pits	79
Crown Colliery, Warmley	81
Goldney Pit	86
Syston Common Colliery	87
The Harry Stoke Drift Mine	89
4. The Mangotsfield Collieries	**94**
Church Farm Colliery, Mangotsfield	97
Parkfield Colliery	100
Brandy Bottom Colliery	102
Shortwood Colliery	105
New Engine and Orchard Pits	109
5. The Collieries of the Golden Valley	**110**
6. Early Collieries at Coalpit Heath and Westerleigh	**118**
Frog Lane Colliery at Coalpit Heath	127
Glossary	**138**

 # Foreword

The Coalfield known as the Bristol Coalfield is divided into two portions. The northern portion covers the City of Bristol and South Gloucestershire, while the southern portion includes Radstock and Midsomer Norton and Somerset. This book only covers the northern portion of the coalfield.

The Bristol Coalfield is different to many other coalfields of the U K. It is a complex basin with the coal measures rising towards the surface on all sides, but in the southern section outcrops are largely concealed by later formations. The measures are mostly steeply inclined and heavily faulted, and occupy a number of detached or partly detached basins.

The southern area of the Northern portion has many coal outcrops running eastwards from Easton to Bitton giving rise to large areas of shallow coal with ancient workings in the form of crop workings and bell pits with drainage levels. The area around Easton is covered by deposits of Mercia Mudstone previously known as Keuper Marl. Unfortunately the southern portion occupies a major part of the City of Bristol and the area of Kingswood, once part of the Forest of Kingswood. In this area a very large number of coal seam outcrop, some of which are now unknown.

Mining appears to have commenced early all over the coalfield with no particular pattern. Wherever there was a demand for coal and coal was present at shallow depth, it was worked, usually for limeburning and smithying and some for domestic use. There is an abundance of documentary evidence for early mining, particularly at Coalpit Heath, which dates from the 1690s and other material ranging from the medieval period down to the 18th century.

Readers may ask why there are no footnotes or references in this book. This is because all of the material in the book is drawn from the archives of Bristol Coal Mining Archives Limited and the collection of the author. The only exceptions are the maps of His Grace the Duke of Beaufort and the list of workmen at the Two Mile Hill Colliery. His Grace kindly allowed the maps and the Two Mile Hill material to be reproduced.

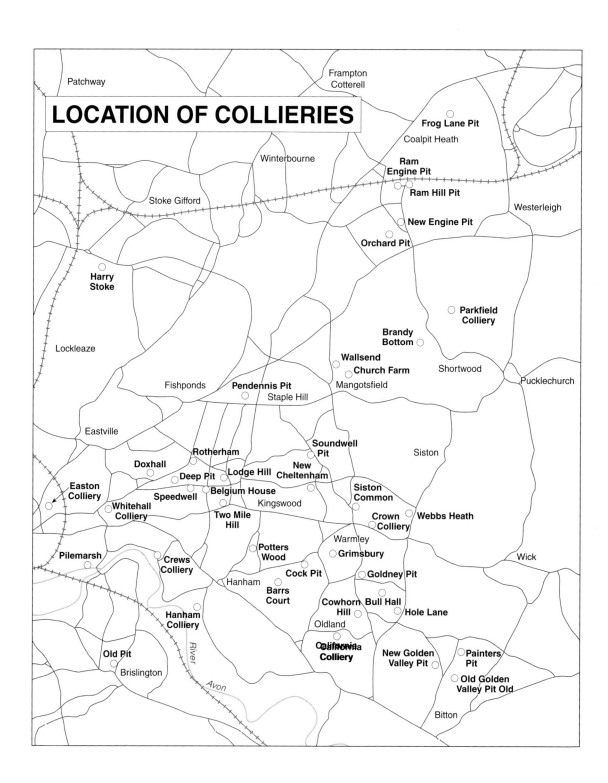

The Coalfield

Mining in the 17th, 18th and early 19th centuries

The above period of mining mainly took place in areas which are now a major part of the City of Bristol and the heavily built up area of Kingswood. Many mining areas which were extensively worked, are now completely forgotten. Motorists driving along the A420 and passing through Two Mile Hill can be forgiven for not realizing that they are passing through an area of coal mining which was first worked in the late 17th century and deep mining here only ceased in 1926 when Hanham Colliery closed. Almost every form of mining had been carried out at Two Mile Hill: crop workings, bell pits, shallow shafts, deep shafts, water levels and pumping from deep workings by Newcomen Engines. Today every trace of mining has been totally obliterated by Victorian and early 20th century development.

In many areas traces of early crop works are still present and are frequently uncovered in site investigations, some as little as only one and a quarter miles from the center of Bristol. Other areas such as Frampton Cotterell, Mangotsfield, Soundwell, Warmley, Cadbury Heath, Longwell Green and Whitehall all have extensive areas of 17th century crop working and bell pits, which are occasionally uncovered in site investigations and building sites. A common feature to all of these areas are the water levels, some of them shallow and some deep and unknown to all but a few. Many were driven in the late 17th and early 18th centuries and many are still carrying water after 300 years. As mentioned in the section on the Newcomen Engine, this coalfield has a number of engine houses (some of them very early), which date from the 18th century. Even now in the 21st century, engine houses are still unfortunately being demolished, as at Crown Colliery at Warmley.

Crop workings are often uncovered on sites and are often not recognized, as they are usually only 10 to 15 feet deep, and are little more than small linear quarries. These are always backfilled, often very little coal is present, but broken sandstones and mudstones with disturbed clays and small quantities of coal dust are always present. Most of the time these ancient works are now well compacted, but occasionally some crop works can produce areas of soft ground. One area where slight remains of crop workings can still be seen is the northern section of the playing field between Gladstone Road and Church Road at Soundwell. Good early remains of crop workings and bell pits can also be seen on Webbs Heath.

One early mine plan dated 1695 in the Record Office at Taunton shows long headings with square shafts every 100 to 150 feet. Most headings were straight although some meandered following good coal or avoiding washouts (areas devoid of coal), but no working areas are shown extending out from these main underground roads.

In Kingswood, Coalpit Heath and Westerleigh and other areas, extensive open workings have been uncovered. Contrary to popular belief, very early workings were irregular. Pillar and Stall workings mainly dated from the late 18th and 19th centuries in this coalfield: undulating seams, faulting and washouts make methodical working on the pillar and stall system almost impossible. Most concerns were abandoned because of poor conditions, disturbed and distorted seams and not the lack of coal. These problems certainly applied to Cock Road. At Warmley Colliery a report of 1841 states that: "The seam of coal usually two feet two inches thick, occasionally expands for a short distance into a swell or gout of 20 yards. One of these extraordinary lumps continued for 100 yards in length".

Often in early workings, a meandering heading was driven through the coal extracting it where it was thick. In many old workings no system was present. Where fragments of old workings have been uncovered they are as described, see page 12, and it was only in the late 18th and 19th centuries that Pillar and Stall workings appear in this area.

One site which revealed very early workings was the section of the Ring Road under and just north of the Hanham Road the (A431), formerly in the Liberty of Sir John Newton. Here the thick Buff Vein was present and was removed from the deep cutting intended for the ring road by contractors. As the coal was excavated workings dating from the 1670 period were uncovered.

Here the Buff Vein ranged in section from six feet to eight feet or more. As the surface area was stripped a number of bell pits came to light. A map in the British Museum dated 1672 shows a number of coal pits at work precisely where the bell pits were found.

Under these bell pits a heading was also found, 20 feet of which was visible, the heading was filled with small coal, obviously thrown back into the old workings once the large coal had been removed. This heading was driven along the strike of the seam. A number of oak pit props were found still in place and when removed were found to be identical to wooden props in use today apart from the fact that they were made of oak which had been split by wedges into sections of 6 x 5 inches. The arms (vertical props) were notched, but no collars were recovered. The vertical arms were all five feet long, but the length of the collars was not known as the southern side of the heading had been removed taking out the collars. The heading was five feet high and had horizontal lagging behind the arms. Brushwood was packed behind the lagging to prevent crushed or small coal from spilling into the roadway. See page 13.

From opencast workings in other areas, it is apparent that only 30 to 40 per cent of a given area of coal would have been extracted from pre-1750 workings. The same amount of coal appears to have been left at Hanham. Later in the late 18th and 19th centuries reworking of pillars took place, but prior to 1750 no reworking appears to have taken place except for areas of Coalpit Heath. This was mentioned in a notebook of 1794. The risk of working up to old workings was always great. Flooding of workings from very early workings occurred in Two Mile Hill in 1735, Lodge Hill in the 1830s and 1840s and Soundwell Colliery in 1853 (which closed the pit). The last colliery to close due to flooding of the workings was California Colliery in 1904. Unfortunately the plans do not show where the water burst in. It would seem that no records were made of the early workings over land around the edges of the California Colliery workings. Some years ago I saw a bell pit in the Cadbury Heath district in which the bottom had collapsed into very old shallow workings, which when viewed from the bottom of the bell pit, revealed open workings extending down dip, beyond the range of a powerful light. These workings, which were in the Rag Vein, were very extensive and were certainly the largest open workings I have ever seen in this coalfield. Probably connected with the California workings.

Unfortunately no open cast working has ever been undertaken in this area, so very few early workings have been exposed, although plans were drawn up in the late 1940s to work some areas of Siston Hill, Goose Green and land east of Webbs Heath. Here shallow coal three to four feet thick was found. These sites which had been given names such as Beatrix, Bella, Clara and Chippy, were unfortunately not worked. Another site adjacent to the Midland Railway at Pomphrey Hill, which had been given the name of Marie, was included in the list of possible opencast sites. It is now realised that these potential sites had been the scene of early mining operations of the 1690-1730 period.

It is known from site investigations that workings of the 1670-1700 period in this area were around 90 to 100 feet in depth, apart from small areas drained by water levels. Early winding was carried out by a simple windlass, similar to that on a well. One illustration exists of a hand operated windlass over a shaft 168 feet deep at Crews Hole in the 1750s. Cog and rung gins were introduced early in the 18th century. With the cog and rung gin, the horse walked round the entire appliance including the shaft. The mechanism was similar to that of a windmill.

Later in the 18th century the whim gin came into use. In this design the rope drum pivoted vertically and the rope was brought away from the shaft by pulleys. For the fist time the actual winding operation was removed from the pit mouth. In some pits particularly in the North of England winding from depths of 400 to 500 feet were achieved using four horses on a single gin. Many Bristol pits were also winding by horse power on shafts 400 feet or more in depth.

At the present time no tools from an early coal pit have been recovered in the Bristol Coalfield, possibly because no opencast operations have been carried out, and the late 19th and early 20th century workings were too deep. Early maps of Kingswood show numerous small iron works who provided tools and equipment for the pits, such as Isaac Smith's Works at Warmley, the Broad Arrow Works at Hopewell Hill, Cools Smiths Shop and the Owls Head Works, both near to Cock Road. Later in the 19th century larger sophisticated Iron Works like Gregory's were established first at Warmley then later on at Kingswood Hill. This firm made large winding and pumping engines for many of the pits. There were also firms like Phipps at New Cheltenham who made boilers, and Rogers at Redfield who made boilers, hudges and kibbles and other equipment.

It is not known when underground rails were brought into use in this coalfield, but the report by the Children Employment Commission of 1841 shows a young boy pulling a sledge along a wooden ladderway or railway This system was certainly in use at the Cowhorn Pit in the 1830s and 1840s. Some pits would have been using a cast iron plateway by this time, however. Other pits were using small five inch diameter cast-iron wheels on an underground plateway. Some of these were recovered from under the heapstead at the Golden Valley Old Pit excavations at Bitton.

Most underground hauling would have been carried by boys using the girdle, a piece of circular rope, with an iron hook, running between the legs to hitch onto the chain of the sledge or cart. The same 1842 report states that in some pits loads would be dragged through passages only three feet high, and some pits would be several inches lower. Few pits used ponies underground at this time. Although it is on record that the Coalpit Heath Collieries had twelve horses at work in 1840, underground horse gins were also used on underground inclines as at Brandy Bottom colliery in the 1840s.

One innovation which was in use in this coalfield early in the 19th century was the sliding trap or stage over the mouth of the shaft. This was where the top-man, standing on the pit top, slid a stage over the mouth of the shaft. He then lowered the hudge onto the stage which covered the shaft and prevented any over-reaching by workmen, or lumps of coal falling back down the shaft into the mine. Early records are full of shaft accidents where men overreached and fell down the shaft, or stepped backwards into the shaft. Stout plaited rope or hempen strap was widely used in preference to chain, iron chain being liable to snap in frosty weather. In the Bristol coalfield there was a preference for flat rope.Headframes were of wooden construction and raged from small simple headframes like Mangotsfield and Crown Colliery, Warmley with one sheave, to large complex structures of Speedwell and Deep Pit with twin sheaves 13 feet in diameter and 30 feet high. Later at Speedwell Colliery a steel headframe replaced the old wooden structure. This was the only one in the northern portion of the coalfield.

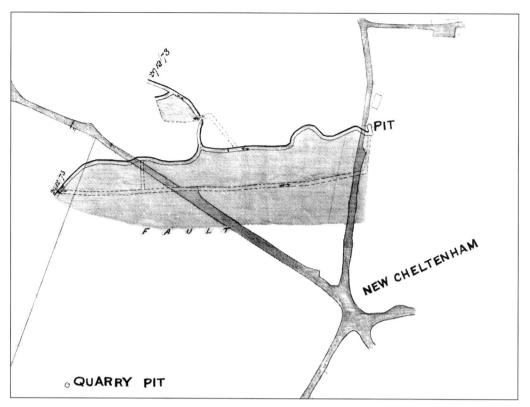

New Cheltenham Pit. The abandonment plan showing the limited extent of the workings when abandoned in 1873.

A coal seam outcrop at Warmley. The seam was often degraded on the surface, and not worth working. The good workable coal was normally found 15 to 20ft down and was always extracted by bell pits.

A bell pit at Magpie Bottom, the diameter of the shaft is around 5ft and the depth of the coal will be 20-30ft.

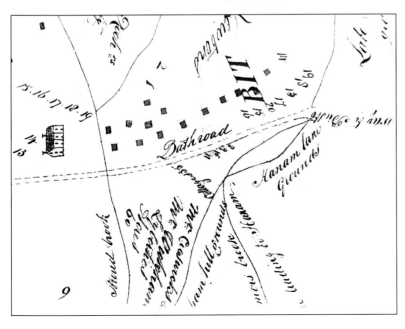

A map of 1672 showing bell pits in the vicinity of Magpie Bottom. The Bath Road is the main Hanham Road. These bell pits were uncovered by a developer.

The uncovering of an early coal work at Longwell Green. This open heading was found at a depth of 15 to 20 feet, and will date from the 1680-1720 period.

An early open working uncovered in Longwell Green dating from the 1680-1720 period. The height of the workings excluding the water was about 3ft.

Above: Oak timbering from the 17th century found in excavations in the Buff vein at Hanham in the early 1980s.

Left: An unlined mid-18th century shaft found at Whitehall in the 1990s. This shaft is sunk through Redcliffe Sandstone which will be about 40ft thick to work a 5ft thick coal seam.

Above: The photograph shows a bell pit from the 1670-1700 period alongside the mid-18th century shaft. This bell pit is a minimum of 40ft in depth. There was no sign of any timbering in either shaft, although it is difficult to see how a shaft sunk through soft Triassic material could function without timbering.

Left: The engine house of Thompson Pit at Cock Road. The structure will date from around 1820. The slot for the flywheel is just visible on the left-hand corner and is filled with brickwork. The window, although old, is not original. The engine was a rotative beam engine with a large flywheel and external winding drum. This engine also drove pump rods at night when not winding coal.

The shaft at Thompson Pit. The shaft was divided into two compartments, one for winding and one the pumping equipment. The drift behind the ladder is the ventilation drift and connected with a ventilation furnace. The rear section was the upcast into the furnace.

The seating for the bell-crank of the pump. This seating is situated on the side of the shaft of the Thompson Pit. A horizontal rod was driven by a crank on the beam engine which actuated the bell crank, which in turn operated the vertical rods in the shaft.

The original wrought iron gates of Thompson Pit. Similar gates were present some years ago at Crown Colliery, Warmley.

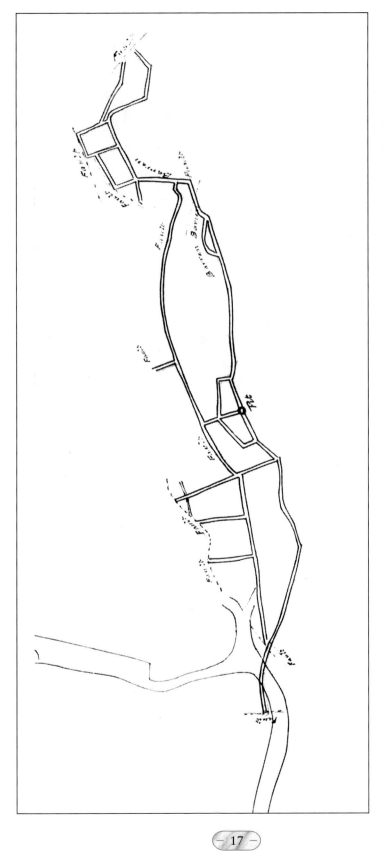

Plan of the Hollyguest Pit also known as Barrs Court Pit. The reason for the abandonment is obvious. The author has never seen such a number of faults in such a small area of working.

An early drainage level on the eastern end of Cock Road which most likely dates from the late 17th century.

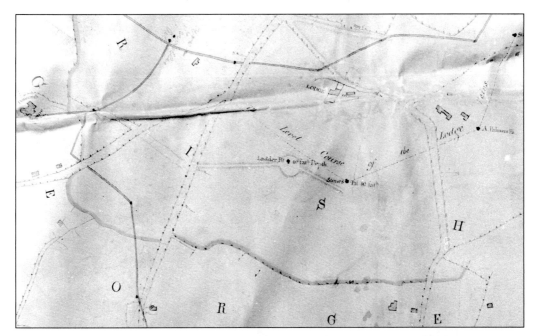

Lodge Hill in the 1803 period. Only five pits were working. The problem was that there were many abandoned and flooded pits which had been working the shallower seams. This curtailed the working of the remaining coal deposits. When the Brain family took over the area from the Duke of Beaufort there were a number of inundations.

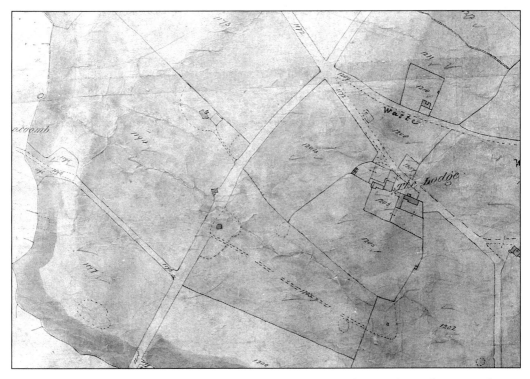

A map of Lodge Hill in 1830. Only one pit, the Lodge Pit is now working, with a coaling pit on the east of Charlton Road and the pumping engine to the north west. It was this pit that sadly had major accidents in its last 20 years of working under the Brain family.

The uncovering of Stone's Pit 420ft deep on Lodge Hill in 1990. A typical shaft of the mid-to late 18th century, ready to be capped. The shaft walling was taken down to bedrock and the capping will be placed on the bedrock.

The water levels of the Kingswood and Fishponds area

The earliest recorded water level is shown, on a map of 1672, in the valley now known as the Coombe Brook Valley. This level was then known as the New Level, so was there even an older level? Handel Cossham stated that this level cut the fault (Great White Fault) and drained the country at a depth of 15 fathoms or 90 feet. This level would have drained part of Lodge Hill which was worked for coal at this time. Cossham could not date the level but suggested that it was well over two hundred years old in 1875. This level was dry and did not discharge any water for many years, but in the late 1970s, it commenced discharging a large volume of thick, highly polluted water containing iron pyrite and other products of oxidation which still flows today. This orange discharge only occurs in old flooded workings which contain some air space, the oxygen reacts on the iron pyrites in the coal seams creating ochreous deposits.

When the coal had been worked out a second level was driven from a lower point from Roe Yate, by the Duke of Beaufort, almost under the now disused viaduct of the Clifton Extension Line. This level drained the hill to a depth of 33 fathoms or 198 feet. Before reaching the Great White-faced fault it branched off in two directions, one to the north and Hillfields, and the second to the middle portion of the district (Two Mile Hill). Around 1790 in the driving of a level near the Alcove, Fishponds, older workings were encountered at a depth of 90 feet. These old workings must have been driven at least 70 years previously, to have been forgotten.

Some of the longest levels were driven from the Siston Brook, The Two Players Levels situated north and south of the main London Road were each two miles in length, and driven between 1700 and 1750. The western section of Players southern level is known as Potters Level and obviously drained pits at Potters Wood. The northern Players Level discharged into the Siston Brook at Tennis Court Road. This level carried water, from the playing field south of Gladstone Street at Soundwell, two miles down to Warmley.

Cools Level at Magpie Bottom drained an area of mining which dated from the 1690-1730 period. To the north two more levels drained the area around and west of Lower and Middle Pits of Soundwell. Many other levels exist: one shallow level was uncovered draining the eastern end of Cock Road some years ago, when the ring road was under construction. This level which was full of water, very nearly flooded a number of new homes when it was breached by a machine when removing soft coal from the intended Wraxall Road underpass. Levels shallow and deep are present at Coalpit Heath and are still active today. An early level was found on the ring road at Hanham and when cut into, it drained water from a mine shaft half a mile to the north west, and 60 feet higher. This level is known to date from the 1680-1700 period.

These early miners were capable of driving through fairly hard sandstones and shales for long distances.

Mine Lighting

Candles were the only source of mine lighting in the Bristol Coalfield. They were carried in bunches and made locally in Kingswood by a number of candle makers. By the late 18th century the candle holder came into use. This T-shaped iron holder, sometimes also made in brass or copper, could be carried in a hat or cap or the spike could be stuck into the coalface or a timber. The last man to use a candle in the coalfield worked at Camerton Colliery in the Somerset portion of the coalfield and only ceased work in 1950, well into his late 70s.

The candle was originally made from tallow, but by the 1860s some candles were made from paraffin wax, which gave better and cleaner light.

Following the candles came the Scotch Lamp and the Peg and Ball Lamps which were also used in the gas free levels in South Wales. These small lamps were usually made of brass, around

A candle holder and two Scotch lamps from the Bristol Coalfield.

A peg and ball lamp used in Frog Lane Colliery.

three inches high and were worn in caps or placed on the floor of the coal face. After World War I some of the spouts of the Scotch lamps were often constructed from bullet cases. Most lamps were made by tin smiths or men at the pits. These lamps ran on tallow, and some are known to have been made in the USA after 1910-20s.

The Peg & Ball lamps used a globular vessel to hold tallow and a wick. It was usually made from iron or sometimes brass. The bottom of the vessel has a circular projection soldered on and this has a screw-thread cut on it to receive the holder when the lamp is in use. It can be carried in the hand or worn in the cap. The Peg & Ball lamp first appeared in the 1880s and continued until displaced by the carbide lamp.

The latter which is still used today by some cavers on the Mendip Hills, gives an intense light, and ran on water and carbide which gives off a highly inflammable gas which could last for four hours on a charge of water. Two charges of carbide and water could last for around eight hours, or a shift. The carbide lamp consisted of a water chamber placed over the carbide container. A valve allowed water to drip onto the carbide which then produced acetylene gas which when ignited burns with an intense white flame.

Early lamps called the Auto-lite were imported from the USA. Later firms like Premier of Leeds and P & H of Birmingham produced a British made lamp. Some carbide lamps in use in the Bristol Coalfield were made by local tin-smiths from the knobs of brass bedsteads. Later large carbide hand lamps were used for areas like the pit bottom, shaft work and in gas free mines which applied to East Bristol and Kingswood They were used by officials.

The Davey or safety oil lamps were not widely used as the coalfield was generally free from gas, but some of the big pits kept several lamps as a precaution and for checking for gas which did occur in some pits such as Easton, Speedwell Harry Stoke and others which were officially gas free.

The Newcomen Engine

It was the introduction of the Newcomen engine into the coalfield which brought about a rapid expansion of the mines. Until steam pumping was introduced, the average depth of a coal pit was unlikely to have exceeded 100 feet. The exception would have been the Kingswood Lodge area, which had free drainage in the form of Levels.

The Newcomen engine was a fairly complicated engine which worked on the creation of a vacuum under the cylinder which then drew the piston and piston rod down into the cylinder, thus lifting the pump rod. The steam was generated at atmospheric pressure in a haystack boiler and filled the cylinder during the upward stroke of the piston and rocking beam. The steam valve was then shut and cold water condensed the steam, which created a vacuum under the piston. The atmospheric pressure then forced the piston down raising the pump rod in the shaft. This was known as the working stroke of an engine. The piston is raised again overbalancing the weight of the pump rods. The Newcomen engine could work at eight to fifteen strokes per minute.

The early engines situated in the Midlands and other parts of the country could only raise water from a depth of 200–250 feet, but when the first engines were erected in the Bristol coalfield from 1739 onwards, they were raising water from depths of 500 feet (Soundwell Colliery). A shaft at Hanham is now known to be over 400 feet deep. Here a pumping engine was certainly working in 1769 (Donn's Map).

The limiting factor of these early engines was the boiler, which did not generate sufficient steam to supply the cylinders. A further factor in the improving efficiency of the engine was the size of the boilers which were slowly increasing. At the end of the Newcomen period, one haystack boiler at the Old Pit, Bitton, was 20 feet in diameter, although this boiler was 19th century in date. Greater precision in the manufacture of cylinders also created a high degree of vacuum below the piston, thus making the engine much more efficient.

Many of the atmospheric engines in the Bristol Coalfield had long lives. The Bitton engine worked from the 1820s until closure in 1898. An engine at South Liberty Colliery worked from the 1760s continuously until it was dismantled in 1900. Brain refers to the Warmley Engine which worked for nearly one hundred years working at two strokes a minute. Unfortunately he did not give a precise location. However there is a strong possibility that Brain was referring to the engine on Siston Hill where the engine house still stands.

The South Liberty Colliery, Ashton Vale, (where the engine is unlikely to be after 1760), raised water from a shaft 700 feet in depth. The engine had a normal working speed of ten strokes a minute, and worked with its original beam until the end in 1900. Several engine houses still stand in this coalfield and are still occupied. One particular engine house stands in Siston, this engine house is shown on Isaac Taylors map of 1777 with a working engine. The interesting feature here is that the beam was situated in the south or side of the engine house, and not as in later engine houses in the gable end.

There is a rather striking engine house in Hollywood Road, Brislington which is early, but unfortunately nothing is known about the history, apart from the fact that it was shown in a painting of 1826 as disused. It is possible that this structure housed a Newcomen engine, as the chimney situated on the rear of the building suggests an internal boiler under the cylinder, which is a feature of early engine houses. The beam- or bob-arch on this structure is situated in the gable end. An internal inspection of this house showed that the recesses for the spring beams are very high up in the building, again suggesting an engine over a boiler.

The well preserved engine house at Siston is probably the finest known early engine house in the coalfield. It is a large building with the opening for the beam situated on the southern or side of the structure, and not the gable end. The recesses for the spring beams in the bed room are far lower that those in the Brislington house. This suggests that the Siston Engine had a side boiler. An arch on the northern end of the house may be associated with a boiler as it gives the impression that it allowed access to the ash pit under the boiler. Recently a second Newcomen engine with a side boiler has come to light at Coalpit Heath. This engine is clearly shown with a boiler house on a map dated 1772.

There were probably many atmospheric or whimsey winding engines which were in use in the Bristol Coalfield. The term whimsey occurs on a number of old colliery sites and certainly refers to the presence of an atmospheric winding engine. One example is the Upper and Lower Whimsey Pits at Coalpit Heath and the Whimsey pit at Soundwell. However, the term whimsey was also used for a horse gin and refers to an engine, which could be horse or steam powered.

Only on two sites have any remains of Newcomen Engines been found. On the site of the Golden Valley Old Pit at Bitton, remains of an atmospheric engine have been found. A windbore was also found in roadwork's on the Ring road at Warmley. The windbore was made by Jones & Co of Cheese Lane, Bristol in 1777. In the excavations of the pumping engine house at the Old Pit, Bitton, site a brass tap was found, although it cannot be certain that this is the tap controlling the water supply to the open top cylinder, there is a chance that it could be. There was an extremely worn brass bearing also found in the engine house. Alongside the top of the shaft, one cast iron link which formed part of the chain which connected the beam to the wooden pump rods was found. A similar broken link was found on the site of the Middle Engine Pit at Nailsea.

The earliest reference to a Newcomen or Fire Engine are the accounts of Lodge and the Fire Engine coalworks for 1739 and in 1748, is this the Old Engine in Hillfields Avenue.

An engine house for a Newcomen engine on south side of Argyle Road, Chester Park

In November 1993 the site of an early engine house which is thought to have housed a Newcomen Pumping Engine, was investigated by industrial archaeologists. The engine house formerly stood on the northern side of the clay pit of the Hollychrome Brick Works, now a council owned playing field at Chester Park. The early history of this site is obscure, but a map dated 1672 shows several coal works and a water level in the area.

In 1754 a lease was granted by the Chester family to Norborne Berkely for all coal works in Kingswood. Norborne Berkely clearly held leases before 1754 in the Kingswood area, as there is an agreement by him and Charles Arthur for the management of the Lodge Colliery, which had been at work since c. 1700.

The earliest record of an engine on this site, is on the plan of the Parish of Stapleton dated 1782. The site is now known as Argyle Road. It is shown as abandoned, as is a coal pit 170 feet to the north. The date of erection of the engine is not known, but is likely to be of the Newcomen type. It could have been erected by either Norborne Berkely, or the Duke of Beaufort after 1775 when all of the Kingswood Collieries were leased to the 5th Duke of Beaufort by the Dowager Duchess. The two pits stood on a piece of ground known as the

Engine Ground on later maps. In 1820 a plan shows the same engine with additions to the engine house and an even later plan of 1830 shows a large reservoir.

Geology

The Kingswood Great Vein is approximately 250 feet in depth in the vicinity of the shaft with the Toad vein at about 300 feet deep. The shaft could have been sunk to either seam and the engine was capable of dewatering either of them.

The excavations

These excavations proved to be disappointing. There was a limit on the depth of excavation as we were in a public place Unfortunately we could not examine the deep area of the boiler and firebox, and most of the bob wall had been removed. There was an opening from the east side of the engine house which allowed access into the stoke hole and boiler. The quarrying operations for the Clay Pit in the early years of the century had destroyed much of the structure and removed the good building stone. Only fragments of the foundations remained, and road works in Argyle Road had totally demolished the rear of the structure.

The exercise did prove that a Newcomen pump had existed for almost 50 years and had been kept in steam until the 1830s, when flooding closed this pit. The engine probably assisted the Duncombe Engine 850 feet to the south west in draining workings which were abandoned in the early 19th century.

At sometime external boilers were erected along side the house, which needed additional water. This site is most likely to have been the Lodge Engine Pit which is incorrectly shown on the geological map as been situated on the site of the Warwick Arms on the corner of Charlton Road and King Johns Road. This shaft is thought to be an air pit.

Developers did uncover the site of another engine which may have been of the Newcomen type on the site of the Goldney Pit, now known as Roy King Gardens. Here a bob wall and large cavity which was probably a cistern were uncovered but destroyed very quickly. The author was shown a photograph taken by the developers after the structure had been removed.

Foundations of a third unknown Newcomen engine was found in a development in Pleasant Road, Downend. Here the tunnel into the ash pit of the boiler and fragments of the engine house walls were uncovered and not recognised, but very good detailing on a site drawing revealed this unknown engine. On a fourth site at Staple Hill the tunnel to the ash pit of the Newcomen engine was uncovered in the site investigations on the site of the New Level Engine, back in the 1980s.

Known Newcomen Engines in Kingswood, Bristol area

Engine house at Pleasant Road, Downend. Found in site investigations.
Golden Valley Old Pit, Bitton. Uncovered in excavations.
The Old Brislington Pit, Brislington Village. Still stands; inhabited.
Corn Horn Hill. Warmley. Documentary evidence.
Conham Copper Works, Conham. Documentary evidence.
Coalpit Heath. 3 engines.
Serrage Engine. Remains and documentary evidence.
Ram Engine. Remains and documentary evidence.
Old Engine pit. Shown on Donn's map of 1769.
Duncombe Pit. Engine shown on map of 1805.
Fudges pit, Speedwell. Old abandoned pumping engine shown on a map of 1805, probably a Newcomen type.

Two Engines, Hanham. Donn's Map 1769.
Two engines at Engine Bottom, Pucklechurch. Donn's Map 1769.
Lodge Engine Pit. Excavated. The engine stood in Argyle Road, Chester Park.
Old Engine. Hillfields Avenue. 1780 documents & map and 1805 map.
A second engine is shown on Donn's map of 1769, well to the north east of the Hillfields Avenue engine.
New Level Engine, Staple Hill. Documentary evidence. Proved in site investigations.
Nibley Colliery, Nibley. Donn's Map 1769 and documents.
Pilemarsh Pits. One possibly two engines. Pilemarsh. Maps and Documents.
St George's, Chester Engine. Documents, Fire Engine public house named after engine.
Staple Hill. Sheppards Engine. Donn's Map
Two Engines at Lower Soundwell Colliery, Chiphouse Road. Documents.
Syston Common Engine. House still stands; inhabited.
Warmley Engine. Mentioned in Brain's *History of Kingswood*, position not clear.
Warmley Brass Works.
Two Mile Hill Engine, Two mile Hill. Donn's map 1769 and documents.
Crews Hole. A sketch by R. Angerstein in 1754 shows what must be an unknown atmospheric engine on the hillside opposite the Beehive Inn.
Rogers book on the Newcomen Engine does list four more engines for which there is no known location. Doubtless more will turn up as additional sites are developed.

This engine house dates from the 1760-70 period as is shown on the map by Isaac Taylor of 1777. It is one of the few structures in the area built from copper slag blocks.

This house is almost unique as the beam was situated in the side wall of the house and not the gable end. The only other engine house with such a feature in the country is at Elsecar in Yorkshire.

The structure has two arches on the right hand side of the house, which have yet to be identified, but are highly likely to be associated with the boiler. Originally the engine would have had a haystack boiler situated under the cylinder but when engines were reboiled, replacement wagon boilers were often installed alongside the house. At Killingworth Moor Colliery in 1762 extra boilers were also situated alongside the engine house although a haystack boiler was placed under the cylinder of the engine. It is therefore unclear whether the arches on the above engine house are original or a later addition for later boilers.

The beam would have been wooden. Unfortunately the diameter of the cylinder is unknown. It is possible that this engine went out of use when Syston Common Colliery was sunk around 1800. Bore holes put down for an intended open cast site on Siston Common in the 1940s, and situated 600 feet to the north of the engine house, (now alongside the ring road), show very extensive workings which must date from the mid 18th century. These workings were shown to be still open, although some were full of water. Some seams were 2ft 6in thick although one was found to be 5ft 6in thick. Probably these workings would have been drained by the pumping engine as they were at a higher horizon than the pumping shaft. Large voids were found at a depth of 66 feet, and these seams would have cut the pumping shaft at a depth of 200-250 feet. The engine was capable and may have raised water from a depth of 400 feet. On the 1904 O S Map the reservoir for the pumping was still present.

Left: The Engine House of the Brislington Old Pit taken on the rear side showing the chimney. The position of the chimney suggests that it was an atmospheric engine.

Below: The interior of the bob arch on the Brislington Old Pit. It is thought that this pit had an extremely short life as it was shown in a painting as disused and in a poor state of repair in the 1820s.

Above: The brick boiler seating for the atmospheric pumping engine at the Golden Valley Old Pit, Bitton. This is late seating and would date from about 1820.

Left: The wooden bob on the South Liberty Newcomen Engine. The date of erection is not known but must have been around 1760. This interesting view shows wrought iron bracing on the king post. Note the tip of the catch pin over the spring beam.

The chain on the outdoor end of the beam bob. The photographs above and below opposite are from *Engineering* Magazine of 1895.

Above: Easton Colliery upcast shaft.

Left: An 18th century shaft found on the site of the Ladder Factory. Staple Hill. In the foreground of this photograph is the slot of the balance bob of the Newcomen engine which was situated behind the camera.

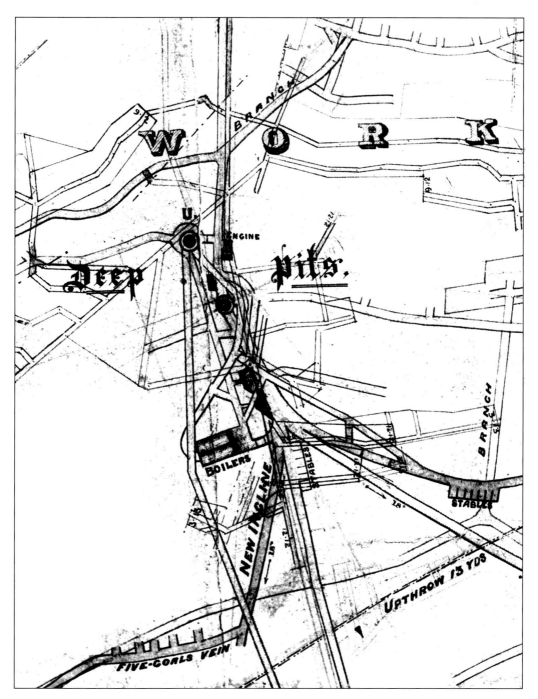

A plan of Deep Pit around 1870, showing the pit bottom area with boilers, haulage engines and stables for horses.

The Western Pits

Easton, Whitehall and Hanham Collieries

Easton and Whitehall Collieries were the most westerly Collieries in the East Bristol District, the colliery known as Pennywell was later included in Easton Colliery and is 450 yards west of the Easton shafts. For many years Pennywell Colliery was a separate concern working the same seams as Easton, and owned by a number of owners. In 1854 the concern was owned by

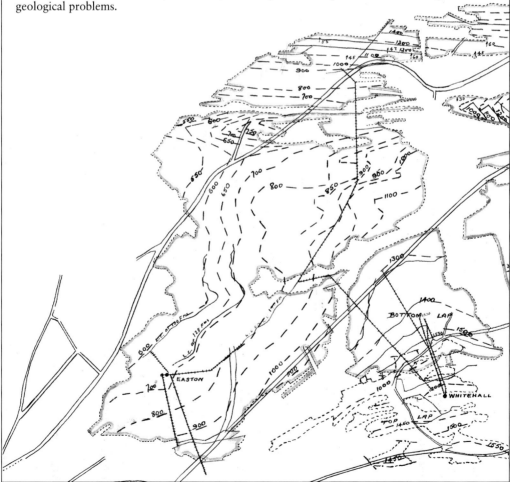

The extent of the workings in the Kingswood Great Vein at Easton and Whitehall Pits. The coal measures of these collieries dip to the south at 1 in 3 with the main workings lying south of the Kingswood anticline. Apart from minors rolls and faults, the Easton workings are fairly regular, except for the ridge in the north where the coal is vertical. The workings of Whitehall Colliery to the south of the shaft are very disturbed, with only the north part of the take being free from geological problems.

Chick & Keeling, later becoming Chick & Brown & Co. In 1866 the owner was Charles Brown changing to Brown and Harris in 1866, but in 1874 the owner was G.W. Harris. Finally in 1877 the concern became known as the Pennywell Colliery Co, until purchased by Easton Pit in 1886.

Easton Colliery was the last of a number of collieries which worked in the area. Coal pits are known to have worked at Clay Hill as early as the 1670s. The first phase of working was around 1660-80, being the working of the outcrops and bell pits at Clay Hill and the area now known as Greenbank Cemetery. These workings were sunk through thin beds of Triassic material. The two Clay Hill Pits and Old Easton Colliery were also sunk through the Triassic material.

Eastern Colliery was sunk around 1830 by Davidson and Walters. Leonard, Betts and Boult had an adjacent pit which closed later. This may have been the large Colliery known as Old Easton Pit in Hinton Road, which is shown as complete, but it is not known whether the pit was still working in 1841. A number of other pits are known to have worked in the area to the east of Easton Colliery in the 1800-1840 period. The two firms united as the Easton Coal Co., and afterwards continued trading as Leonard, Bolt and Co. becoming a limited company in 1880. Easton colliery in 1841 employed 160 hands and only three boys under 13 years of age, although one was seven. The youngest boys only made 2 shillings a week. The other carting boys would earn from 6 shillings to 10 shillings a week.

The Brickyard pit was 126 fathoms or 756 feet deep, the other pit only 80 fathoms or 480 feet. The seam at the former (The Great Vein) is four feet thick, and at the latter two feet ten inches. The Two pumping engines were respectively 40 horse and 80 horse power.

Easton Pit had a covered hutch of plate iron that could carry seven men which was needed in a shaft much troubled by runs of water.

Easton Colliery taken around 1900. The low structures in the foreground housed the seven Lancashire boilers which drove the steam winder and fan. The tall engine house held the vertical winding engine which had been made by Gregory of Kingswood Hill.

The Easton Colliery consisted of two shafts 30 yards apart. Twenty yards of New Red sandstone and marls were passed through at the top. These marls were later used in the brick yard also on the colliery site. The down cast shaft was 11ft in diameter having been enlarged in 1858-59. The shaft was walled throughout with 9in brickwork to a depth of 1,080 feet. The upcast was oval, to a 7ft by 5ft, walled-in like manner. It was deepened to a depth of 1,920 feet. This deepening proved the 'Parker' group of seams and the Ashton seams.

The sections in the Easton Upcast of the Ashton group of seams are as follows:-

	Thickness	Depth from surface
Top seam	2ft 0in	1,710ft
Good coal	3ft 0in	
Great Seam in three divisions	7ft 0in	1,800ft
Little seam	4ft 0in	1,920ft

A report written in 1891 states that "the probability of there being 16ft of workable coal is the Lower or Ashton Series, enough mineral exists to extend the duration of Easton Colliery for eighty years". However this coal was never fully worked.

The winding engine at Easton was made by Gregory of Kingswood in 1858. The engine had two vertical cylinders 26in in diameter, and a 5ft stroke with an overhead cylindrical drum 14ft in diameter, for round ropes made from crucible steel with a 30 tons breaking strain. The engine could raise 1 ton of coal per minute.

The pumping engine was of the Cornish type, made by Bush of Bristol in 1838, of 90 h.p. The cylinder was 67in diameter, 9ft stroke in cylinder and shaft. This replaced the original cylinder taken out in 1883, the weight of which was 6.35 tons. The water was raised from the sump to the surface. The engine made five strokes per minute, working fifteen out of twenty four hours a day, delivering 250 gallons per minute.

A Waddle fan, 25ft in diameter, was erected on the upcast shaft at Easton, driven by an horizontal engine 20in cylinder, 3ft stroke; direct acting, producing a circulation of 35,000 cubic feet of air per minute. When demolition work was undertaken on a structure built over the upcast shaft, the fan drift connecting the shaft to the Waddle fan was uncovered and the top of the shaft was also uncovered and found to be brick built. A smaller Waddle Fan had also been installed at Pennywell Colliery.

At the Whitehall Pit the waste heat from the boilers near the upcast shaft and the exhaust steam from the hauling engines together produced a circulation of air through the pit amounting to 20,000 cubic feet per minute. In the late 1880s the steam plant at Easton was served by seven Lancashire boilers, each with double furnaces. The winding engine was also fitted with an automatic non-overwind device to cut off steam from the winding engine; it also applied steam to a brake engine. An indicator was fitted to show the cage position within 1in; these devices had been in continued use for 25 years in 1883.

Easton had its own basket shop, making baskets mounted on steel runners for use underground. They would have contained about 1 cwt each and would have been hauled by a lad with a tugger. Each boy (13 years old) served two men, hauling about 2 to 3 tons per day. However the pay was about three times as much as that for surface work. The boy would haul the basket to the foot of a steep 60 yard incline, where he tipped it into a tram, which was hauled up by a surface engine through an endless rope running through the shaft. From the top of the incline horses hauled the trams to the pit bottom. The horses were watered from a spring 40 fathoms (240ft) down and were bedded on sawdust from the sawmill, which kept their coats in good condition. One seam in Easton was 16 feet thick, but contained a dirt band and rubbish between the beds. This seam was difficult to timber. At the head of the incline driven down to the Ashton Seams was an old beam engine which was used as a haulage engine. This engine was the original steam winding engine from the shaft.

Whitehall Colliery

When the work begun on the Whitehall shaft in 1860, the sinkers and Leonard Boult & Co did not realise that the site of the intended shaft was over an area of extensive workings dating from the early to mid 18th century, which ranged in depth from the surface down to 300 feet. Only 850 feet to the north-east was a disused shaft and horse gin house. Other shafts were present just north of the Whitehall Road and may have been visible. There was also a map available dated 1769 which showed a coal pit on or near to the abandoned horse gin, and only 800 feet from the Whitehall shaft. Some years ago, further old shafts dating from the 18th century and possibly earlier were found in the Rose Green Playing field, yet more evidence of extensive early mining. Back in the 1980s when Rose Green school was demolished two more shafts came to light with depths of 40 to 50 feet and 60 feet deep. These shafts were clearly part of the complex of early workings. It was not surprising that the sinkers struck old water logged workings.

The 1860 shaft was sunk with a diameter of 13 feet, and was 1,140 feet in depth. The shaft was divided into two by a brattice of 9in brickwork, the area of each division being 3 to 1 for the downcast and upcast respectively. A travelling way was made between Easton and Whitehall Collieries in the event of an accident at one or the other pits. Double doors in the travelling way made the ventilation of the two mines separate. The Whitehall shaft was sunk from the surface, to the depth of 300 feet, through old workings which were heavily watered. These workings are now known to be of mid 18th century date and were the cause of the subsidence in the old Co-op Bakery. The sinking of the Whitehall shaft was completed in 1867. Because of the presence of old workings the lining of this portion of the shaft was made exceptionally strong, with 3ft of firebrick, built on a wooden wedging crib as a foundation. The bricks were set in mortar composed of lias lime and ashes from the boilers, the grouting behind it consisted of small stones cemented in the same kind of material. A series of small wooden pipes were placed in the grouting, the whole depth, giving passage for the escape of gas and water. The shaft below the old workings was lined in 9in brickwork set in hydraulic mortar.

The winding engine at Whitehall Colliery, the house of which still stands today, had two 30in horizontal cylinders with a 5ft stroke together with a plain 15ft diameter drum which raised coal from a depth of 1,140 feet. The engine was built by Gregory of Kingswood in 1860. This engine also wound water in tanks situated under the cages. This operation was only required for a few hours, twice in a week. There were two hauling engines placed near the bottom of the shaft with two Lancashire boilers 30ft by 7ft diameter to supply steam.

By 1880 the method of working the different coal seams was by longwall, the gateways or hatchings are 25 yards apart on an average, each level 60 yards apart. The excavated spaces are filled up entirely by the stone obtained from heightening the places above the seam. A self-acting incline is erected in a suitable position, the coal from several stalls on each side being sent down this incline. The Mueselar Lamps were in use in the Parker seam and the Ashton group of seams in Easton. All other workings had naked lights in the form of candles. The colliery only worked the Doxall Vein and the Kingswood Great Vein, the Great Vein being cut at a depth of 700 yards.

Pilemarsh Collieries

The Pilemarsh pits were sunk some time in the late 18th century as separate concerns, although it is not known who owned which pit. Mr Butler was one owner and Samuel Riddle another. In 1799 water from an old abandoned pit broke into one of the Pilemarsh Pits and drowned four men. This pit was owned by Mr Butler. By 1803 they are shown as one concern with three large steam engines, which are shown with boilers situated alongside the engine houses. It was most likely that these engines were Newcomen Engines. The Pilemarsh Pits comprised of six shafts at Pilemarsh and two more 1,000 feet north of the Blackswarth Lead Works. They are all sunk on or close to the outcrops of the six coal seams of the Upper Series. Two of the engines were situated on the western side of the Netham Road and the third, 300 feet west of the lead works. This third engine is possibly situated over a shallow drainage level which appears to drain into the river Avon. It must be that the Pilemarsh pits were sunk and worked to provide coal for the Brass Works furnaces and the Blackswarth Lead Works.

A section of the Old Pilemarsh Pit shows that the following seams were worked:

Francombe Seam at 18 feet
Upper Millgrit Seam at 180 feet
Lower Millgrit Seam at 210 feet
Rag Seam at 240 feet
Devils Seam at 480 feet
Buff Seam at 498 feet

The main coal seam here was the Buff Vein which, although variable, could be six feet thick and was said to be the most productive seam. The Parrot vein which proved to be a thin but high quality coal in great demand by the Brass industry did not prove at Pilemarsh. The Millgrit Seam is shown in another section to be in two bands each 2ft 6in thick, where as the Rag Seam is only 1ft 6in thick . No thickness is shown for the Devils seam which may not have been worked at Pilemarsh.

The Old Pilemarsh Pit is situated east of Netham Road and certainly had a steam pumping engine. Although no documents exist, the positions of a number of shafts east of the Netham Road does suggest a shallow water level discharging into the River Avon. An accident was reported in 1798 when a child, George Cantle, aged 5 was scalded and burnt to death, when he fell into a pan of water belonging to the fire engine, which belonged to Messrs Butler. In the next year four miners were drowned in the Pilemarsh Coal Pit when water burst into the pit, once again the presence of old workings appears to have been forgotten. There is a total absence of mine plans in the 18th century in this coalfield, although many maps of shaft positions only are still in existence. After a number of different owners, Pilemarsh Pit ceased working in the late 1850s.

Opposite: The winding engine house at Whitehall Colliery which also held another Gregory engine. This was later than Easton and was a horizontal wider with two cylinders.

The Great Western Colliery

This colliery was sunk in 1847 by the Great Western Coal Company but due to various geological problems the workings were abandoned in 1857. The shaft was sunk to a depth of 901 feet but they only worked one seam at a depth of 582 feet. A document in the archives of Bristol Coalmining Archives Ltd by Moses Rennolds of Bedminster, dated 1862, states that

'The seam varied from 6 to 12 inches, was very faulty, and in some places produced no coal at all. The workings were continued by the Company for some time with the hope that the seam might be found better ahead, but such did not prove to be the case and the working was discontinued and a serious loss was sustained by the Company'.

The Great Western Colliery was abandoned in 1857, but in the previous year a project was commenced driving a heading 6ft by 5ft northwards at a depth in excess of 700 feet, towards the bed of the Feeder Canal, but was stopped when the heading was 3ft short of it. A letter dated 25th June 1856, to J Harford of the Bristol Docks Board mentions that if the permission is not granted to continue driving northward, under the land of the Bristol Rock Board, closure of the colliery will result. A further letter from the Bristol Docks Engineers Office of the 1st November 1856 still did resolve the matter. After a further letter of the 3rd November 1856, no further correspondence occurs, and we can assume the project was abandoned. In 1861, correspondence shows that the owners of the site were owed seven years rent, and were trying sell or rent the abandoned colliery. There was even talk of the lessees permitting agents to descend the shaft for the purpose of ascertaining the state of the works. Clearly the winding engine and boilers were thought to be in working order. In the Chancery documents held by Bristol Coalming Archives Ltd and dated 1862, it is suggested that "the filling up and levelling of the pit should take place". But from a visit to the site before the construction of the new Link Road, many buildings were still standing around the heapstead and the shaft, which was capped, but was not filled.

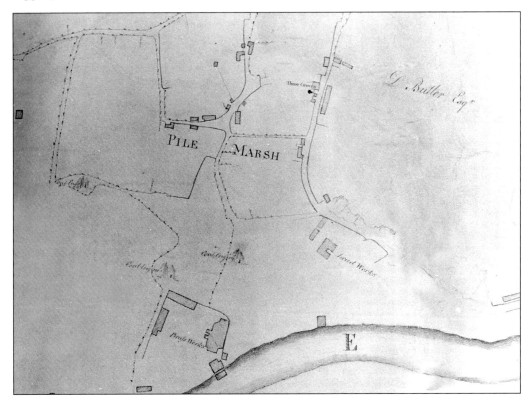

Above: The interior of the heapstead of the Great Western Colliery in the 1980s. The shaft which has never been filled is under the wooden frame to the right of the picture.

Opposite page: A map of 1805 showing the three Newcomen pumping engines which were at work at this time. A fourth pit is shown with a horse gin, adjacent to the Three Cocks Inn, which is still in business in Blacksworth Road, although rebuilt in the early years of the 20th century.

Mines in Crews Hole

The early years of coal mining in this area are not well documented, but it was certainly the shallow coal deposits which led to the presence of the copper and lead works in the Avon Valley and the establishment of Copper works at Conham in 1696 and Crews Hole in the 1700-10 period. A sketch by Reinhold Angerstein in 1754 shows the Crews Hole valley looking eastwards; two hand windlass can be seen on the hillside high up the slope and behind a steam pumping engine. Angerstein gives the depth of the coal pits with windlasses at 168 feet. This would put the shallow pits on the Upper and Millgrit Veins and about 300 feet to the north-east of Beehive Terrace, on the western end of Crews Hole Road. A Glass kiln is also visible which is presumably the kiln situated at Strawberry Lane.

There are a number of shafts in the Crews Hole Valley which are thought to date from the 18th century. More shafts must exist, for deeds and leases exist showing large tracts of land in Crews Hole particularly to the east of Strawberry Lane which had been worked for coal prior to the working of the fireclay deposits. The plans of the Troopers Hill Mine which worked the Buff Vein suggest that one of the levels may have been driven to work the Buff coal seam first.

The only remains to be seen today in Crews Hole are those of the chimney of Crews Hole Pit also known as Troopers Hill Pit. This colliery is not shown on the 1803 map of St George but is shown on the Tithe Map of c.1840. This colliery is shown with a steam engine and a horse gin, and worked with a second shaft 550ft to the north-east. A Loxton drawing of the second or northern engine house, shows that the house is built in the same style as the lower pit with an identical chimney of the early 1800s. A notable feature of the engine house in the Loxton drawing is a large archway set in the side of the structure for installing and changing the cylinder of the engine, which again is a feature of the early 19th century. This pit worked the Millgrit Seam 2ft 6in thick at a depth of 122ft. The Rag vein is also shown on the shaft as variable in thickness and is 180ft in depth. No closure date is known but the colliery had ceased work by 1845.

A remarkable occurrence took place when boreholes were being put down in 1989. At a depth of 80ft, the borehole cut through a standing pit prop in partially open workings of Crews Hole Colliery in the Millgrit vein. This bore hole is nearly 500ft south-west of the shaft.

The first recorded working of fireclay is in the 1850s when the Brain Family were listed as owning and working a fireclay mine at Crews Hole. Unfortunately the location is not known at the present time. The area of land situated to the east of Strawberry Lane was worked first for coal, then later a fireclay mine was opened around 1895-96 working the clays associated with the Nos.1 and 2 Veins, the Devils Vein and the Buff Vein which was cut in a level in the highest point of the mine which is now under the cemetery. The last date was of 1900 on this plan. A second plan of the mine on Strawberry Hill appears to commence around 1900 and worked until 1909, working the Buff Vein clays. In fact the fireclay workings broke into the earlier coal works. The last date on the second plan is 18th April 1910. Do the two plans mean a change of ownership?

The mine situated under Troopers Hill is mentioned in 1878. Earlier records appear not to exist, and it is possible that the mine may have commenced prior to 1878. The owner was a H.C. Burge who also managed the mine. In 1884 the mine was worked under the Bristol Fireclay Mine Co. Ltd, with Mr Burge as manager. The mine plans show that the Buff Vein clays were extracted. but this mine ceased working in 1908. There was certainly a mine on this site well before 1878, as the abandonment plan shows an old level being cleared. These clearance works were stopped on the 20th May 1908, which is also the last date of working the clay. There was a further fireclay mine 200 feet to the north-west of the old Beehive Inn, but the plans show that this was not an extensive mine and was only 220 feet in length. It was disused around 1901. This mine is close to the position of the fire engine shown in the 1753-55 drawing of Crews Hole.

Plan of the Troopers Hill Mine of the Bristol Fireclay Mine, made on abandonment in 1908. The feature marked 'chimney' is the chimney which still stands on Troopers Hill.

One of the three entrances of the Troopers Hill Fireclay mine uncovered in site investigations made before the construction of a housing estate. The main workings are all situated to the north of the main road and are under Troopers Hill. No workings are under the housing in this area of Crews Hole.

Remains of the chimney of Troopers Hill Colliery, now a listed structure.

 The Kingswood Pits

Two Mile Hill Collieries

The earliest date for the working of coal in Two Mile Hill is a map dated 1672 where several coal pits are shown as working on either side of the main road.

In a news paper report of 1735 the following story is recounted:

'In 1735 three men and a boy were trapped for ten days in a coal pit near to Two Mile Hill. These workmen were workings 234 feet underground, which was very deep for the time, and fifteen years before the first steam pumping engine was introduced.

The states that these men were on the brink of a precipice a further 96 feet in depth. They were wedging out coals on the 7th of November 1735 when a prodigious torrent of water burst out of a vein, and put out their lights.'

The paper stated that:
'crawling on their hands and knees from place to place to avoid the water , they providentially got to rising ground, and reached a hollow place, whence coals had been dug, and there continued. They found a bit of beef and a crust of bread, together about a quarter of a pound, which they divided equally and eat. It was for some time very easy to get water, but the water sinking , it became more difficult, till at length, not being able to obtain it, they were forced to drink their own urine, and to chew some chips which they cut from a basket, but losing their knife, even this miserable expedient failed them, and one endeavoured to eat his shoe.

They were almost suffocated with heat and the nauseous fumes that arose from their bodies, and continued without any other sustenance till the 17th November, when their friends, after several ineffectual attempts, let down a large quantity of burning coals, which dissipated the black vapour, and the water being gone off in a great measure, five men ventured down and calling out, were surprised to find them alive and able to answer.

The eldest was about sixty years old and was delirious, and all of them were very weak and for a time intirely blind. They were taken out, and having received some refreshment, walked to their homes to the great astonishment of a vast crowd of people assembled from all parts.

The men did not apprehend that they had been above five days underground. When the water had burst in upon them, there were four other boys in the mine, but being at the Tip of the work ran to the rope crying to be pulled up, and not withstanding it was done with as much expedition as possible, yet the water was at the heels of the last boy, who, as the other three were hawling (presumably hauling) up, caught hold of the feet of one of his companions, and all got safe out.'

The depths quoted in the paper of 1735, 234 feet and tip of another 96 feet making a total depth of 330 feet, if correct are exceptional. Although earlier workings are shown on the 1672 map, workings of this depth were unexpected.

But no workings of this depth have ever been recorded for the 1670-1700 period. Any drainage level running southwards would have to have been over a mile in length and would have only allowed a depth of 130 feet of free drainage.

What system of drainage allowed workings in the 1670-1700 period to work down to depths in excess of 200 feet. It is possible that these early pits used some form of buckets to raise water, but at the present time it is not known when the cog and rung gin displaced the hand windlass which would have limited the depth of working to about 150 feet.

Rudder in his *History of Gloucestershire* of 1779 makes the comment that the "Two Mile Hill Pits belonging to his grace the Duke of Beaufort are 107 fathoms deep (642 feet deep). At this pit and many others they used fire engines to draw out the water." A map by Maule of 1805 shows a coal pit on the northern side of the main road as Tylers Pit, not to be confused with the Tylers Pit at Speedwell later to be known as Deep Pit.

Information for the 14th January 1771 tells us that 175 men were employed at Two Mile Hill Pit, although four of these men were working at Whites Hill Pit, as hewers and there were three sinkers employed at an unknown pit.

The seams worked were the:
Two Coals Vein, at 2ft 6in thick.
Doxall Vein, at 1ft 6in thick.
Five Coals Vein, at 5ft thick.
Kingswood Great Vein, at 4ft thick.
Two Feet Vein, at 2ft thick.
Gillers Inn Vein, at 1ft 10in thick.

In the first week of January 1771 the colliery employed 54 hewers, i.e. a face worker who cuts coal, plus 78 fearers, (the job description is not clear, but is likely to entail the removal and carrying of coal from the workings faces). There is also a Veerer who is now known as the banksman, he removed the tubs from the shaft and returned tubs back down the shaft. There is some confusion between Fearers and Veerers. There were also 53 deadwork men, who did various jobs, like clearing falls of stone, putting up timber, and they were always paid by the day. Additionally there were 3 branchers, a man employed in driving a branch or making a tunnel in the solid rock.

The accounts also show that 22 dozen candles were purchased for the sum of £7.3.0 (6s 6d per dozen, the price also charged in Derbyshire at about that time [pers. com. Lindsey Porter]).

This figure is not the true number of candles used in a week, as one man would probably burn a dozen candles in a shift so 175 men would need around 175 dozen candles a day, or 1050 dozen candles a week. The hewers were paid from eleven to eight shillings a week, the wages for fearers ranged from around 7 shillings down to five shillings and one or two men were only paid 4s 6d a week. The highest wage paid to a deadworker was 12 shillings although the normal was 10 down to 8 shillings a week.

A map of 1805 by W Maule shows the pumping pit to be situated on the south side of the road with the coaling shaft on the northern side known as Tylers Pit. This is not to be confused with Tylers Pit at Speedwell later to be known as Deep Pit. The map by Donn published in 1769, shows an engine. It was also mentioned in 1779; it would have been of the Newcomen type. The engine lifted water into the drainage level which ran northwards down Charlton Road. A letter dated 1809 asked whether George Baylis should put up the Two Mile Hill Engine. It would appear that the concern was in decline and must have ceased work soon after.

The Two Mile Hill Pit of the Duke of Beaufort was not the only concern working coal in the late 18th century. There were other smaller pits in Two Mile Hill known as Barratts which were only 650ft south-east of the Duke's Engine pit. Some 1000 feet to the south of the Duke's Engine pit was Jones Pit which collapsed in very heavy rain in 1988. Just to the east of Jones Pit were three more pits, one unnamed pit and Hudds Pit which is shown to be in the Chester's Liberty on the map of 1790. There was also a pit with a large horse gin house known as Whittucks. There was finally a smaller pit west of Whittucks known as Denglys Pit about which nothing is known apart from the fact that the pit was worked by a horse gin and was in use in1803. As in many old areas, the names of the sinkers or owners remained with the pits however small until they were abandoned and forgotten.

Lodge Hill and Hillfields

This is one of the earliest areas to have been worked extensively, and covers the area now known as Lodge Hill and the Hillfields estate. The earliest date for the working of coal in this area is an agreement dated 1650 between Sir Maurice Berkeley and Nicholas Smart of Mangotsfield, coal miner for mining coal in Kingswood.

In 1656 a Captain John Copley set up an iron-works at Hopewell Hill just to the east of Kingswood Lodge. This iron-works was built next to a number of coal pits situated close to the main reservoir at the Soundwell Road works of the Bristol Water Company. This iron-works was intending to use pit coal for smelting, using some form of bellows which may have been driven by a windmill, but after spending many hundreds of pounds the project was abandoned. Some years ago in the 1960s, when the main reservoir was under construction, remains of a circular structure came to light with much small coal and clinker; perhaps this was the remains of Captain Copley's iron-works.

There is a coal seam known as the Hole or Hard seam, present at a depth of 60ft which would have been worked at that period. However there is also no mention of any coking of any form in Dud Dudley's account of the works in 1665. Had Copley understood the importance of the coking process, the iron-works may have been successful. The venture was unsuccessful, but it does show that coal mining was long established by the 1650s, and may even date from the later years of the 16th century.

A second group of leases dated 1687-1702 for the Lodge Colliery gives details of veins etc., the yearly rent and a certain sum for every 20s 0d worth of coal dug. At the same time, articles of partnership (dated 1693) were drawn up between John Thomas, John Sutton and William Symes. There was also a dispute between John Berkeley and the partners for not constructing a level to drain the coalworks according to the terms of the lease of 1699. A drainage adit known as the New Level is shown as working on a map of 1672; it drained the area now known as Chester Park and Lodge Hill and is also mentioned in the papers of 1709. It was presumably this level which was mentioned in the dispute with Sir John Berkeley and it may be that the driving of this level was not completed and that the area around Kingswood Lodge was left without free drainage for some time. In the 1980s this level suddenly became active again, discharging polluted water into the Combe Brook. The accounts for the Lodge Coalworks for the year 1709 give a list of the workmen employed in the colliery; the various workmen are as follows:

Hewers, in the Great Vein	4
Carters, in the Great Vein	4
Vearers in the Great vein	2
Landsmen, in the Great Vein	4
Hewers in the Five Coals Vein	4
Caters in Five Coals seam	4
Fearers in Five Coals	5
Landsmen, in Five Coals seam	4
Hewers, in Thurfer Seam	2
Carters, in Thurfer Seam	2
Vearers, in Thurfer seam	2
Landsman, in Thurfer seam	1
Tipsmen	4

In addition there were sinkers; a cartmaker; 2 smiths and 3 engine drivers. They were driving a horse whim and not a steam engine. Hewers were mostly paid 5s 0d per week, carters 4s 0d per week with vearers 4s 0d per week, tipsmen 5s 6d to 4s 0d per week and Landsmen 5s 0d per week. Sixty workmen were on the books of Lodge Coalwork for the second week of April

1710. In that week 2,650 carts of coal were landed, but sadly at this time the amount or weight of coal in a cart is not known. A cart is a very archaic term for a basket, but the capacity at Kingswood is unknown.

Surprisingly in the weekly accounts of Lodge and Fire Engine coalworks for 1739 what must be a Newcomen Engine is first mentioned. In 1748 the month's accounts again refer to the Fire Engine. This must be the earliest recorded Newcomen engine in the Bristol Coalfield, and is possibly the engine shown on all early maps as the Old Engine which is in what is now known as Hillfields Avenue.

There was a lease in 1754 by the Chester family to Norborne Berkeley for all coalworks in Kingswood. In 1775 there was a memorandum of agreement between the 4th Duchess and 5th Duke for the lease of all the Kingswood Collieries for £500 p.a. for the remainder of the term of the lease from the Chester family to Norborne Berkeley (who had then been enobled as Lord Botetourt). In 1787 all of the coalworks on Lodge Hill which were owned by the Chester family were leased to the Duke of Beaufort. The collieries finally became the property of the 5th Duke on the death of the Dowager Duchess in 1799, but before then thirty-one of the coal pits in the area of Lodge Hill were already shown to be abandoned. Only five shafts east of Thickett Road, and two shafts later known as Duncombe pits were left working. These seven pits were clearly powered by horse gins. The shafts alongside the two steam pumps were still maintained.

There are two pits on the Hillfields Estate situated in Sumerleaze which are known as Great and Little Dicot. Both are late 18th century pits 420ft and 360ft in depth and both were sunk to the Little Toad Vein, and were abandoned by 1820. A map dated 1803 shows just two coal pits on the crest of Lodge Hill, the Ladder Pit 40 fathoms (240 feet) in depth and Stones Pit also 40 fathoms deep. The later shaft was uncovered in the 1990s to the south-west of Cossham Hospital, and was found to be square with rounded corners typical of the 1760s period. Stones and Ladder Pits were old 18th century shafts which were reopened and may have been deepened, as was the Stable Pit at the bottom of Clare Road, which is recorded as 40 fathoms (240 feet) deep; this shaft collapsed in 1997.

A further nine shafts are shown on the 1803 map as working in an area north of Woodland Way. The pumping engine on the southern part of what is now Hillfields Avenue was situated on a shaft 40 fathoms deep and raised water to a water level which ran south and then turned eastwards eventually running into the Combe Brook at Royate Hill. The date for driving this level is not known. There was a second pumping engine which stood in Argyle Road which is described in the section on the Newcomen Engine. The engine was possibly disused in the 1780s, and was probably brought back into service and worked into the early 19th century, as a map of 1830 shows a substantial engine house which had been enlarged with a large square reservoir.

As the coal reserves became depleted after 140 years of working, the later workings had to exploit to deeper seams. Lodge Colliery which was the last pit to work in the area was working the Kingswood Great Vein 4ft 6in thick at a depth of over 500ft and the Little Toad Vein 1ft 1in in section was proved at a depth of 630ft, although it is not known how much of this thin seam was worked. An even deeper pit was the aptly named Deep Pit which was 678ft deep. This pit was situated in the north part of Hillfields Avenue and is known to have been at work in the 1789-1803 period. One Lodge Hill pit which appears not to have been worked by the Duke of Beaufort is Nick Melsom's New Lodge Pit which was at work in 1797. Handel Cossham commented in 1875 that Nick Melsom's Toad vein was probably discovered at the New Lodge Pit. In 1822 the Duke of Beaufort leased Kingswood Colliery to Aaron, William and Moses Brain and Francis Faulkener. Aaron Brain then lived at Kingswood Lodge. Twenty-two years later the Duke then leased Kingswood Colliery (also known as Rotherham Pit) to John Monks. Unfortunately the name Kingswood Colliery was used for Speedwell Pit and various other pits in the Lodge Hill area which has caused some degree of confusion.

There were many accidents in the Lodge Hill Pits over the years. An incident occurred in 1833, when water from old workings broke into the mine and trapped five boys aged 13 to 15 without food for six days and nights. Fortunately the boys were rescued unhurt. Six years later in 1839, flood water from another abandoned Lodge Hill pit broke through a barrier of coal which was being robbed and eleven men were drowned.

A report in the *Bristol Mirror* of Saturday June 1st 1839 states:

'that had the water broken in an hour earlier, the loss of life would have been much greater, many of the men and boys having left the pit to take their dinners.

Upon personal inquiry at the pit on Wednesday afternoon, we find that the water has gained on the engine, which has been incessantly at work since the accident, 120 feet of water was in the shaft, the pit being of course completely filled, no hope of rescuing the sufferers is left. The melancholy occurrence has spread the deepest gloom in the neighbourhood, and it is not expected that the pit will again be worked. The workmen had broken into an old pit which had been closed for nearly 100 years, and the first intimation of danger was the extinction of the candles from foul air, which preceded the water. The men nearest the pit mouth rushed to the rope, and clambered up in a cluster, two of them in a state of nudity.'

Apparently a similar accident occurred in the same pit (which was also known as the Causeway Pit) about 1829; then only one young man lost his life. After the 1839 inundation, the pit continued in production for some years.

In the report on Employment of Children in 1841 the Kingswood Lodge Coal Works were described as employing 96 boys and young persons aged from 9 years to 17 years of age. It is not clear whether it was this pit which had been inundated with water in 1839. Lodge Colliery was shown on a plan of 1845 as intact and possibly working, but was certainly abandoned just after 1845.

The pumping engine mentioned in the 1839 disaster is the engine house in Argyle Road. The coaling pit was probably situated 500ft to the south-east on Charlton Road, where the 6in geological map shows what is called the Lodge Engine Pit. However, between the two pits is the Great White Fault with a massive downthrow to the west of 480ft; the geology is indeed complex. In 1873 John Anstie, writing on the *Coalfields of Gloucestershire and Somerset*, says of Lodge Hill:

'That coal has existed here in abundance is quite evident, from the immense quantity of shallow pits, in some places to the number of twelve or fifteen in a single field, as at Kingswood Lodge. Some few are of greater depth, as 100 or 150 yards. From the irregular way in which it has been worked over, there may be much coal left in this district but the danger of meeting with water from old workings is too great a risk to be run for the chance of success.'

But Handel Cossham writing in 1875 disputed this statement, and gives a far lower number of shafts per field on Lodge Hill than John Anstie. This is not quite the end of the mining legacy on Lodge Hill. Now that the area is being developed, shafts and shallow workings are still being uncovered and are having to be made safe and stabilised. The geology of the Lodge Hill area is complex and still not fully understood even today. The Hill is bounded on the west by the Great White Fault which is present under the eastern section of Argyle Road and at one time was visible in the eastern end of the clay pit, now filled. A large unnamed fault runs east-west almost up Ingleside Road and cuts across the junction of Soundwell Road and Downend Road. The area is situated south of the Kingswood Anticline and the coal seams here should dip to the south, but so far every seam found in excavations and various forms of works dips to the north. One can only assume that the two faults have tilted up the seams, reversing their dip.

At the present time four coal seams have been found to outcrop under Lodge House and the site of the old Lodge Hill Reservoir. What is thought to be the outcrop of the Kingswood Great Vein, around 4ft thick, was uncovered close the site of the new Doctor's Surgery which was built to replace the old surgery on the corner of Soundwell Road and Chase Road. This was formerly the surgery and home of Dr Vinter and his wife Dorothy. He was a well known local historian who wrote the first history of the Kingswood Coalfield in 1964. This surgery had also been built over a coal seam outcrop and shallow coal which had been worked in the 18th century. It was these workings which caused the severe subsidence which damaged the rear of the house.

Fishponds and Staple Hill

Both areas are situated over the coal seam outcrops of the Upper series which contain the Cock, Chick and Hen seams which were mainly worked in the 18th and first few years of the 19th centuries. There are several other seams in the Upper series like the Rattlebones seam, a hard coal 2ft thick, which must have been worked although no records now exist. The Stinking seam 1ft 6in thick, was also worked in Fishponds in the 18th century, despite the presence of sulphur, hence the name Stinking vein.

The earliest reference to mining in Fishponds is the name Colepit Lease on a map of the 17th century, the land is shown as lying just west of Channons Hill. The same map shows several coal pits at work to the east of Lodge Causeway in the vicinity of Filwood Road. On the edge of Fishponds there was an early area of coal mining at Clay Bottom and Clay Hill, as two early coal pits are shown as working on the map of 1672; in 1790 two pits here were owned by the Duke of Beaufort; these two pits may be on the same sites as shown on the 1672 map. It is also possible that Clay Hill Colliery was later known as Old Doxall Pit. The later Doxall Colliery dates from the mid 1850s and had two shafts, but the colliery was abandoned in 1865 due to an inrush of water, probably from Old Doxall Colliery. This was one of Handel Cossham's pits. The abandonment plans of Doxall Colliery of 1865 show what appears to be an unknown steam pumping engine just south of the Combe Brook, or very close to the site of Clay Bottom Pit which is on the 1672 and the 1790 maps and also on a map of 1830.

Pendennis Pit around 1907. This photograph probably shows the pit after abandonment, as there is no winding rope. This rare photograph is by courtesy of the Downend History Society.

Further to the north in Fishponds, maps dating from the late 18th century and 1805 show pits on either side of the main road between the Causeway and the Railway Station; both of these coal pits were worked by the Duke of Beaufort. The New Cock Pit was 31 fathoms (186 ft) deep and was recently uncovered when the site of the old Randall's Timber Yard was cleared.

The Hen Pit sunk on the Hen vein was situated on the northern side of the main road. Two early 19th century pits which are now almost forgotten are the Royal Oak Pit in Forest Road and Victoria Pit also in Forest Road, the former was sunk on the anticline and worked the Parkers top seam at a depth of 120 ft. The Victoria Pit, obviously sunk in the early years of Queen Victoria's reign, worked the Little Toad seam at a depth 300 ft, and was sunk on the edge of known 18th century workings. A little known pit was the Star Pit situated on the western side of the Alcove Lido, which was said to have been sunk to the Britton and Stubbs seams. Some limited working of these seam was undertaken, but the pit was soon abandoned some time prior to 1850. Most of the mining in the Staple Hill area also dates from the 18th century again, working the seams of the Upper Series, The pits were mainly worked by the Sheppards and later by Bragg.

There were four steam pumping engines in Staple Hill, one on the site of the Ladder Factory and one in Pleasant Road, Sheppards Engine in Albert Road just north of the old bus depot, and the fourth at the New Level Pit. All were of the Newcomen type. There are sale particulars of a Newcomen engine at Staple Hill Colliery which was sold to the Golden Valley Co, by a Mr Peterson & a Mr Boult, but unfortunately it is impossible to say in which pit the steam engine worked.

The Staple Hill railway tunnel was originally driven as a small tramway tunnel. Old coal workings were encountered and the autumnal rains caused a length of tunnel to collapse in November 1843, so that the whole length of the tunnel was nearly lost. Later in 1843 and 1844 it was enlarged from its original size of 9 ft wide and 12 ft high. The shafts for these old workings have never been found. There were attempts around 1906 to drive a drift mine into the Cock vein, and a shaft was sunk close by Pendennis Road and a drift driven under what now is the Staple Hill Primary School in 1906-07 to the Cock vein. Both ventures were abandoned by 1907 when it was found that most if not all of the coal had been worked out by Bragg about 100 years earlier.

Hanham Colliery

This colliery was certainly winding coal by 1871. The original date of sinking is not clear and is thought to predate the first raising of coal by many years. The colliery was near the River Avon and just off the northern side of Memorial Road; an incline ran from the colliery down to the Avon from where much coal was moved by water. Two shafts were sunk, the downcast was 13ft in diameter, the upcast 6ft 6in in diameter both are lined throughout with 9in brickwork. They were 900 feet

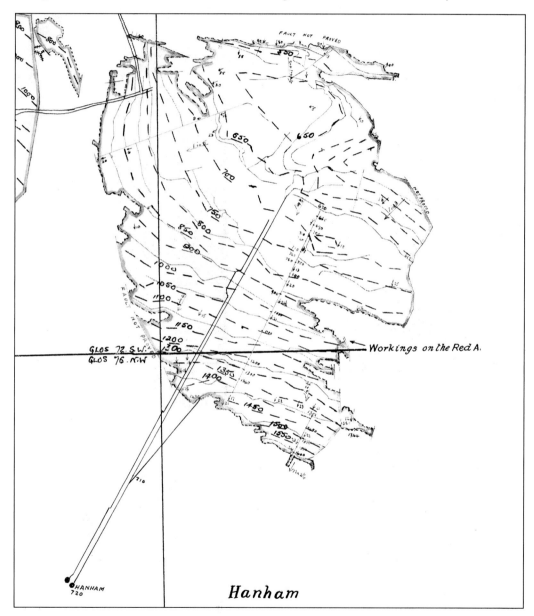

The area of working of Hanham Colliery which was mainly in the Hanham Red Ash vein which was 3½ft thick, the Kingswood Great Vein was displaced by an upthrow fault and was little worked at Hanham. The two inclines dip north east from pit bottom and were a mile in length. The small area of working at the pit bottom was in the Millgrit Vein and the second area of working was the Kingswood Great Vein which did not prove well at Hanham Colliery.

in depth to the Parrot seam. The downcast was stopped short of the Parrot seam, in the hard Pennant Rock above it, in order to drive out the inset and cross measure drift in this rock. The upcast shaft is sunk wholly through the lower beds of the Pennant, its base being the Parrot seam.

The colliery take was large and the workings eventually stopped 500ft north of the main Kingswood Road under Kimberly Street and Broadfield Avenue; the western boundary was under Two Mile Hill Church, and the Great White Fault to the west and south of Two Mile Hill Church, was never breached. To the east the workings terminated on the western end of Courtney Road and Claypool Road. The Hanham winding engine had two 22in horizontal cylinders, a 4ft stroke, and a 9ft plain drum. The load was raised in 45 seconds, two trams holding 9 cwt each being raised on one deck. The pumping engine was a direct acting 'Bull' engine resting on wrought-iron box girders placed over the shaft. The cylinder was 40in in diameter and 9ft stroke; the water was raised in two lifts from the sump of the shaft, the lower one was a bucket lift, 300 feet in height, with a 8in working barrel, and the upper is a forcing set, 150 yards in height with an $8^{1}/_{2}$in ram. This delivered to an adit 150 feet below the surface. This engine made six strokes per minute working at night time only, delivering at the adit 130 gallons per minute. One hauling engine was placed at the surface, the rope running down the upcast shaft. It hauled sets of six trams up a cross-measure incline, 480 feet in length, the dip of the incline was 6in to a yard. From the downcast shaft a cross-measure horizontal drift was made to the south for 1,200 feet, which proved all of the existing seams in the Pennant rock from the Parrot seam to the Millgrit seam.

The coals were thick and of good quality, but the feeders of water presented a serious obstacle, pouring out of the rock in increasing quantity, which caused the workings to be abandoned. The working of coal was then concentrated upon the north side of the shaft, several seams having been cut by a cross-measure horizontal drift, one mile in length, driven in a straight line, with mechanical haulage. Unfortunately an upthrow fault to the north was encountered, near where the Great vein should have been met with, which introduced ground considerably below that seam. A short distance in the Red Ash or Venture No.1 was met with, of $3^{1}/_{2}$ ft thickness. On the discovery of the Great seam and other seams met with in the stone drift, production continued until closure in 1926.

Opposite top: An 18th century shaft at Hanham. This shaft is shown on Donn's Map of 1769 as having a Newcomen Engine. The shaft had never been filled and was only found when the pavement collapsed. It was filled by British Coal in the 1980s.

Opposite bottom: The Hanham Colliery rescue team, with the downcast or coaling shaft in the background.

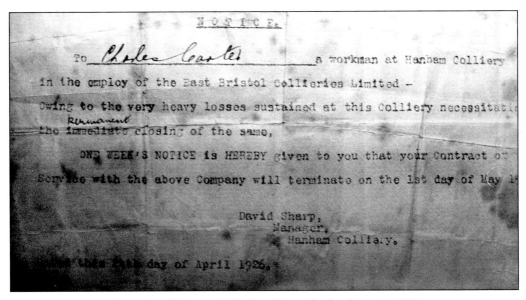

The notice of closure of Hanham Pit to Charles Carter in 1926.

The restored winding house for the horizontal steam winding engine in the 1980s.

Mines in the Cock Road Area

No history of Kingswood would be complete without mentioning the Cock Road area, historically one of the most interesting areas, the home of lawlessness and the Cock Road gang. The area is geologically as complex as its history. By the time that mine plans were being drawn up, the mines in Cock Road had all been abandoned, and very few written records have survived apart from field names and the odd engine house.

The earliest known coal works are in Gages Road where clay tobacco pipes were found on a colliery site where three shafts had been uncovered, these shafts were around 90 feet deep. The pipes were dated to the 1660-70 period, so presumably the pits were of the same date. Other shallow workings and bell pits dating from the same period had been uncovered close to Cock Road Bottom. A map of Gee (More) dated 1755 shows five coal pits workings in an area of 23 acres. The interesting thing about Cock Road is that the names which are mentioned in 18th and early 19th century documents are still around today. Families such as the Brains, Brittons, Cains, Wilmotts, and Webbs, are still well known in the area. A map dating from 1750 shows several sites. The Owls Head Works, and the Cool's Smiths Shop concerns were smithy's or small iron works making tools and equipment for the coal works, of which there were many active in the area at time.

Another area which contained a number of shallow bell pits dating from the 1700 period, was Pettigrove Road. South from Pettigrove Road the seams dip to the south and become complex. The geological map says the "coal measures of the Mount Hill Area are believed to have complex structure". (i.e. heavily faulted). This accounts for the lack of coal works in the area around Hollyguest Road. The only colliery sunk to the south of Hollyguest Road is known as Hollyguest or Barrs Court Pit. It was sunk around 1860-64 by Pilditch, Steedes and Holway to a depth of 740ft, and attempts were made to work a seam 2ft 6in thick, which may have been the New Smiths Coal. This was a total failure, as the area was so heavily faulted, and so heavily watered, it was impossible to work, and was abandoned after working for some three or four years.

To the east of Cock Road conditions were better and the Engine House, which is possibly Thompsons Pit, still stands. This engine house was for a vertical beam winding engine which wound, and also pumped at night when the engine was not winding coal. The shaft was uncovered in the late 1970s and a drift to a ventilation furnace was seen in the side of the shaft. It is said that one old inhabitant remembered the remains of the furnace stack in the next door garden which would have been situated at the eastern end of the furnace drift. It is thought that the engine house dates from around 1820, and copper slag blocks were used as quoins. The pit is shown on a map of 1843 as standing intact and may have been working.

There is a second engine house lower down on Gee Moor about which nothing is now known; it is not clear whether this is a pumping engine house or a winding engine as the house has been surrounded by recent additions and it is impossible to determine its original purpose. It is likely that it was a pumping engine.

Further to the east is a pit known as Smiths Pit about which nothing is known. At Cock Road Bottom, 500ft north of Smiths Pit, stood Cock Road Old Pit which is said to be 978ft deep, and would have been sunk some time in the late 18th century and deepened later. This seems very deep for the Cock Road area. The Old pit is recorded as having worked the Great Seam which is 'very irregular from 0 to 6 feet'. The 2ft seam is also said to have been worked below the Great Vein. Recently, to the west of Cock Road Bottom a number of shallow bell pits and a coal crop were uncovered. Shallow workings dating from 1670-90 are also present on the extreme eastern end of Cock Road and were drained by the water level found when the ring road was under construction.

This is not quite the end of the Cock Road story. Some time around 1890-1900 Hanham Colliery drew up plans to drive through to Cock Road and Thompsons Pit at a depth of 100 fathoms (600ft) and connect them with a branch driven from Thompson Pit bottom to work the Toad Vein, and to drive under the Barrs Court water. Clearly this scheme came to nothing.

The Kingswood Collieries: Speedwell and Deep Pit

Both are shown on a map of 1790. Deep Pit was known as Tylers, just to the north Cottle's pit and was working with a large new steam pump, which was then connected to the Duke of Beaufort's Level. Speedwell Colliery was known as Starveall Pit. A slightly later map dated 1803 still shows Tylers Pit working with a horse gin but Starveall is now known as Starveall Engine, although the horse house gin is still shown. Duncombe and Belgium pits, which later became part of the complex, are shown on a map of 1781 as working with horse gins, but by 1805 Doncom (later known as Donkham) had a large steam engine, which is likely to have been a steam pump. By 1838, Tylers was known as Deep Pit, and the name Starveall was still in use for the later Speedwell Pit. On the 1841 Tithe Map, Starveall Pit has lost its horse gin house and a house for the steam winding engine is the main building on the site.

Rubens Pit was the upcast shaft and was situated some way to the north and it is highly likely to have been a ventilation furnace; the 1841 map shows a small structure on the shaft which may be a furnace. From 1822 onwards the pits were worked by the Brain family, but the workings were confined to the Toad vein series, and the seams above. By the mid-1830s the Kingswood Great Vein Series was discovered, and from then on nearly all of the workings were in the Great Vein Series. By 1841, 300 hands were employed at Deep Pit and Starveall; 123 were boys of all ages, but not more than 30 were under thirteen. A day's work for a boy was hauling 70 carts a day for 1s 2d.

By the mid-1850s Speedwell and Deep Pit had been linked and it is thought that the ventilation furnace was placed near the bottom of Speedwell Pit, where it would have been more efficient.

Speedwell Colliery around 1899 with a wooden headframe. This photograph was taken from where the fire station now stands, looking west. The horse and carts are waiting to load up at the screens situated under the large sloping roof.

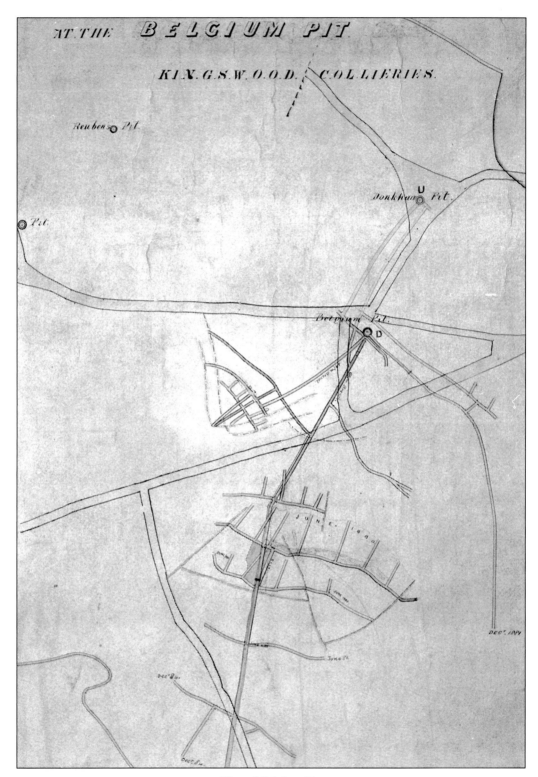

Plan of Belgium Pit.

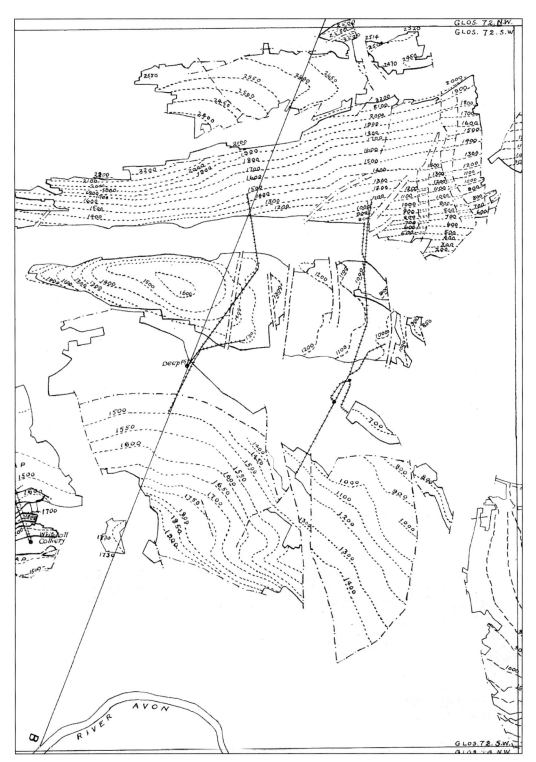

The full extent of the workings of Speedwell and Deep Pit. These workings were bounded on the east by the Great White Fault. The Circular Fault 2000ft to the north of Speedwell Pit divided the workings into two blocks, and the effect of smaller north south faults can be seen, with the effects of the Kingswood anticline.

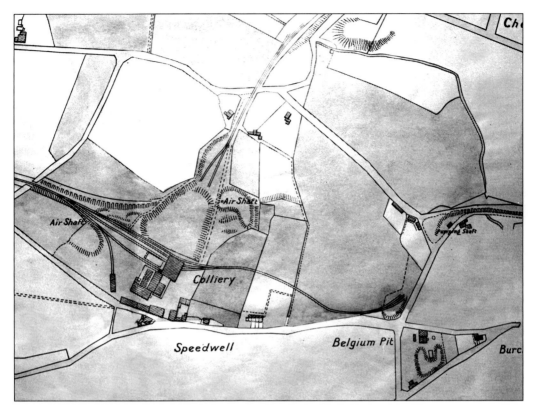

Plan of Speedwell Pit in 1899. The air shaft is known as Rubens Pit which collapsed in the 1970s in the playing field of Speedwell School and was filled and capped by the NCB. The Duncombe Pumping Pit and Belgium Pit are on the left of the plan.

Deep Pit in 1899. Unlike Speedwell Pit the screens are in the open. The locomotive was made by Fox Walker, and the locomotive works were just behind the camera.

Deep Pit looking west in 1899, with five-plank wooden wagons of the Midland Railway.

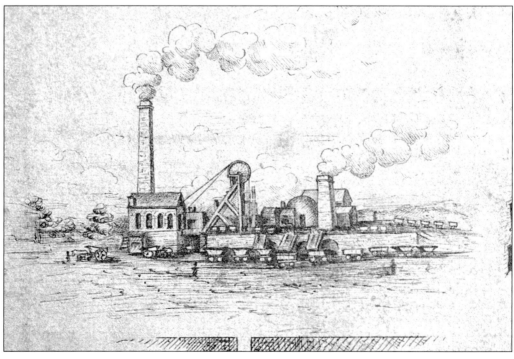

A drawing of a haystack boiler on the second shaft at Deep Pit from a plan of 1880 and always known as the picture plan. It is possible that this boiler drove an engine on the surface which had an endless chain running down the second shaft to the incline which ran northwards.

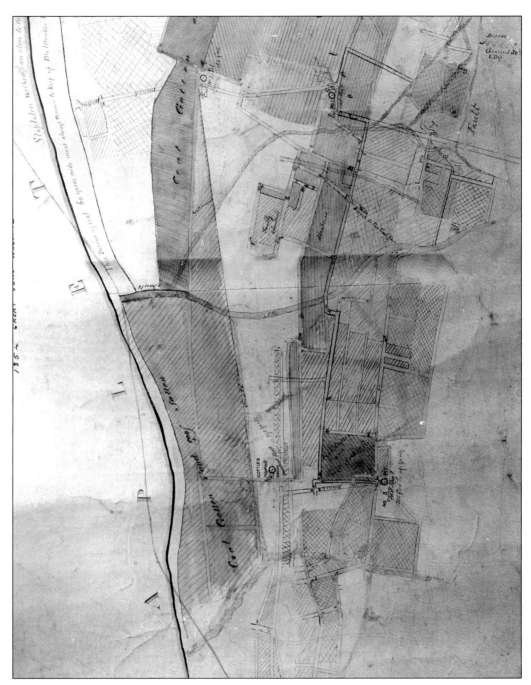

An early plan of Deep Pit in 1854 when under the ownership of the Brain Family. This plan shows the workings in the Kingswood Great Vein which was 4ft thick, the different types of hatching denoting different dates of working. The areas of extraction are pillar and stall workings and not longwall faces. The surveyor has only shown the total area of extraction and has not drawn the detailing of the pillar and stalls. Long wall working was introduced later by Handel Cossham.

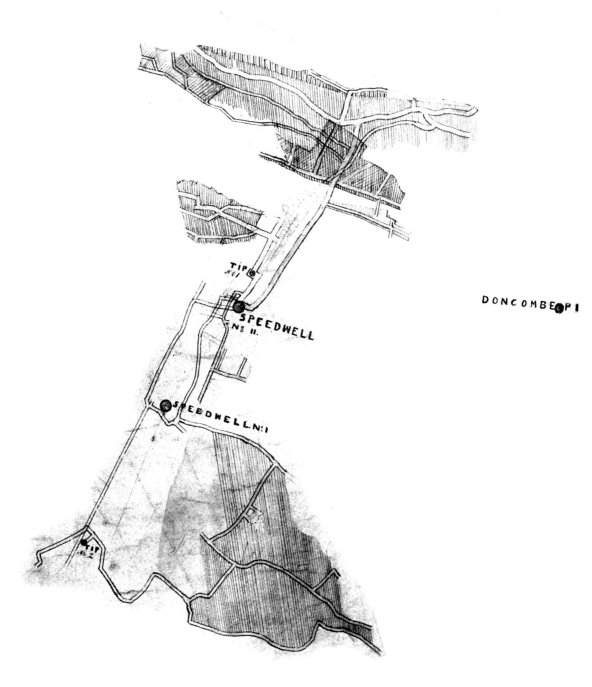

Speedwell Colliery around 1840. This plan shows the workings in the Five Coals seam at an early period. It was opened up after the upper or shallow seams, the Doxall, Toad and Hard seams were exhausted. Note that there is no connection between Speedwell Pit and the pumping pit at Doncombe. At this depth, any water levels and other headings would have been in the higher Doxall and Toad Veins, which would have been heavily watered. The deeper Five Coal Seam was relatively dry.

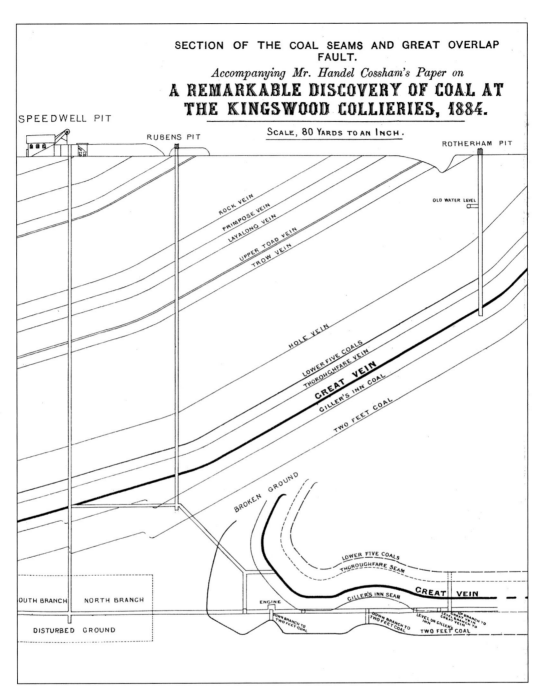

A drawing of the famous overlap fault, which shows the main or lower portion of the Great Vein discovered at Speedwell Colliery.

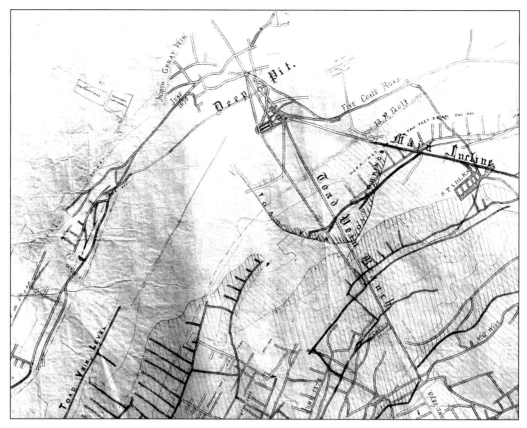

A small section of a large and beautiful plan on linen, probably by Handel Cossham's surveyor, drawn up around 1890-1900 when the collieries were put up for sale after Cossham's death.

In 1850 Kingswood Collieries consisted of 6 shafts:

No.1 Speedwell, upcast, 1,140ft deep, coal winding.
No.2 Rubens, downcast, 960ft deep, now coal winding.
No.3 Deep Pit, downcast, 1,230ft deep, winding.
Donkham Pit, pumping only.
Cottles Engine Shaft, downcast, 870ft deep, pumping.
Doxall shaft, winding only.

A section which is not dated, but is around 1850, shows the Cottle's Engine raising water to the Duke of Beaufort's Level at a depth of 21 fathoms (126ft). The pumps are in four stages: the first are $10^1/_2$in and are of the plunger type, 21 fathoms in depth; the second 11in and 44 fathoms deep; the third 11in and 50 fathoms deep; the fourth were $9^1/_2$in and 30 fathoms in depth. The total depth was 145 fathoms or 870ft. The windbore was offset in a separate lodge or sump.

Deep Pit had at this time a large haystack boiler which was shown on plans of the 1880s. The winding engine also drove an endless haulage system down the shaft to a depth of 145 fathoms or 870ft. Winding at this time in Deep Pit was by a hudge. Deep Pit had at that time two shafts in operation and the haystack boiler was on the second shaft. It would appear the main winding shafts had more modern boilers.

Plans of Speedwell and Deep Pit dated 1854 show the two pits had linked in the Kingswood Great Vein and that Deep Pit, Cottle's and Ruben's shafts were then downcast shafts with Speedwell the upcast. Even in 1854 Speedwell had a set of underground boilers presumably for underground haulage engines. The plan only shows the Great Vein workings with the various types of hatching showing the month or year of working. By 1863 Speedwell and Deep Pits were leased to Wethered, Cossham and Wethered. Later in 1867 the Kingswood Coal and Iron Company was floated by Cossham and Wethered. Handel Cossham bought the mineral rights of the St George area in 1875.

At the same time he also bought the manor and mining rights of the properties of the Duke of Beaufort. In 1879 the Kingswood and Parkfield Colliery Company was formed with Cossham as the main shareholder. Once Handel Cossham assumed full control of the Kingswood Collieries he embarked on a programme of modernisation and additional development.

Belgium Colliery was an old shaft which is shown as an unnamed pit on a plan dated 1790. It was reopened and modernised by Cossham and had an active life of just over 20 years, mainly working the Kingswood Great Vein. The sale particulars of 1900 show that this shaft was by then only used as an air shaft, that had a winding engine with 10in cylinders with 20in stoke. A 7ft drum was geared 3 to 1 and powered by an egg-ended boiler. By the 1880s production was then at its peak. The quantity of coal raised per day was 500 tons; ventilation was by furnace, although a Schiele Fan was under construction in 1885.

The mineral area was large, it extended from Snowdon Road and Manor Road in the north, to St George's Church and Summer Hill Road in the south. The southern area was split by a major north-south fault running through Bell Hill. The bulk of the production came from the Great Vein. It was in great demand not only as a house coal but as a locomotive coal. A large amount was also used at the various gas works in Bristol and Bath.

In 1899 the collieries which included Parkfield employed 1,000 men and over 210,000 tons of coal were raised in the three pits; 131,000 tons came from Deep and Speedwell Pits which by then had been working in excess of 120 years.

Down to the year 1835 the workings were confined to the Toad Vein series and the beds lying above the Toad Vein. It was around this time that the Kingswood Great Vein was discovered, and has been largely worked since. The Toad Vein series had been supposed to be exhausted to the north of Speedwell Pit. A level cross measure drift was driven northwards to prove the Ashton seams, but after driving through 200-300 yards of disturbed strata, some of which was nearly vertical, beds of fine shales and coal seams were found lying almost horizontal with a dip to the west. The manager Mr. Jacob Sparks soon realised that they had not found the Ashton Series but had found the main section of the Great Vein which was to provide another 50 years of fine quality coal. The previously worked area of the Great Vein was the upper section of the overlap, which had been broken off and thrust right over the main part of the Great Vein, which was the main and highest quality seam of coal in the coalfield.

After the death of Handel Cossham the three pits were purchased by a new company The Bedminster, Easton, Kingswood and Parkfield Collieries Limited. The sale particulars of 1900 make interesting reading.

The winding engines at Speedwell, the maker of which is unknown, had 30in cylinders with a 48in stroke, the winding drum was 14ft in diameter and was fitted with a steam brake.

The ventilation was provided by a Capel ventilating fan which had a 7ft inlet. There was also a Schiele fan and the colliery had a total of nine Lancashire boilers.

Deep Pit had a set of horizontal winders with 30in cylinders, 5ft stroke and the winding drum was 16ft in diameter, with four Lancashire boilers.

The Cottles Pumping Engine was a 3 valve Cornish Pumping Engine with a 56in Cylinder, a 7ft 9in stroke with lifting pump 12in diamenter, 65 fathoms deep and forging pump 11in diameter, 35 fathoms deep. Steam was raised by two egg-ended boilers.

The Cottles engine was visited by the late George Watkins around 1918 and he thought it to be disused.

The main pumping engine was the Donkham Engine. The original pump had long been dismantled and by 1900 a vertical direct acting pumping engine with a $26^5/_8$in cylinder with a 5ft stroke raised water 70 fathoms (420ft).

Deep Pit also had an extensive Coke Works that comprised of 24 ovens which were served by a washery engine driven by a single 9in cylinder with an 18in stroke. This washery survived until the early 1950s although most of Deep Pit by then had been demolished. The collieries were both connected to the Midland Railway at a point known as Kingswood Junction by a private line that belonged to the colliery company. This line ran under the washery and alongside the coke ovens. It was also used by nearby Peckett & Co., locomotive builders, under an agreement whereby Pecketts are entitled to send goods over the line, which should not exceed 100 tons in any one day.

A large number of plans still exist drawn up under various owners, but it is interesting to see how the later plans become more accurate, the pre-1850 plans cannot be relied on as they are not precise. Page 62 shows the extraction of the Five Coals seam near to the Speedwell shafts. This plan is not dated but will have been made prior to the link-up in the 1850s. The plan dated 1854, which was drawn up under the Brain management, does show the first underground boilers at Speedwell Colliery which were fed with an air supply from the downcast shaft (Rubens).

A very nice plan of Deep Pit in the 1880s drawn on linen, Page 64 shows the extent of the pre-1830 workings in the Toad Vein, with the stables on the main incline. Also shown are the workings in the Two Feet Vein abandoned in December 1880, directly under the Toad Vein workings.

Under the trust of the will of Handel Cossham MP, the trustees were directed to build and endow a hospital on or in the neighbourhood of Kingswood Hill, for the treatment and relief of sick and injured persons of both sexes, and in order to enable the above trust to be carried into effect the Directors are now offering the Property of the Company for sale.

The collieries of Deep Pit, Speedwell and Parkfield, were purchased by a new company, The Bedminster, Easton, Kingswood and Parkfield Collieries Limited. In 1923 the two Speedwell pits employed 690 men, but by 1930 only 482 men were still working, when the collieries closed in 1935 and 1936, there were only 310 men left working.

Rotherham Colliery

This was situated on the western end of Argyle Road close to a large house called Lindens, which was demolished in the 1970s and now known as Linden Close, on which flats were built. This colliery was first shown on a map of the 1780s as an abandoned pit; it is next shown as a working pit under the name Lodge Pit and was worked by Edward and John Monks, when in 1861 a collier, William Tanner aged 42, was killed by a fall of stone while setting timber. On 7 November 1863 a mason, Isaac Rogers aged 45, died, when 'the end piece was not properly effected, so that the flat wire rope slipped right out of it, and through the rivets, whereby the cage and men fell to the bottom'. The rope should have been doubled at the end, or as it was termed 'turned up'. A collier, I. Lovel, was killed by a fall of stone in a heading on 2 July 1866. This last entry was under Lodge Colliery, although the 1863 pit was also known as Rotherham Colliery. The abandonment plans show that the Great Vein, Five Coals seam and Gillers Inn Vein were worked. Unfortunately the date on these plans has faded and was not legible, but was sometime in the 1860s.

Monks also sank a shaft known as the New Pit in the 1860s, on the north side of Lodge Causeway. Only one heading was driven under the northern end of Russell Road for 200ft and intersected the Toad Vein, Gillers Inn Vein and the Kingswood Great Vein. However little or no coal appears to have been worked. The shaft was sunk to a depth of 642ft.

Soundwell Colliery

Although no sinking dates for the Soundwell Collieries exist, the Lower pit must have been sunk in the late 1740s by Samuel Whittuck. Since parts of a Newcomen engine were delivered and installed in 1750, the engine could have only been at a shaft which had been completed prior to that year.

The Soundwell Collieries consisted of three Pits: the Upper Pit in Gladstone Street, which was 398 yards deep; the Middle Pit also known as the Centre Pit, which was situated at the junction of Gladstone Street; and Church Road (the site now known as St. Stephens Close), the depth of which is said to be the same as the Upper Pit. This pit straddled the road with the shaft on the north side of the road and the winding house on the south. The Lower Pit with two shafts, one for winding and the second shaft for pumping, was situated at the bottom of Chiphouse Road and is recorded as having been 1,080ft deep.

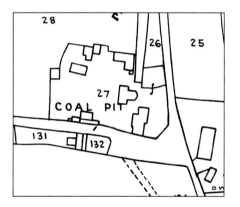

Soundwell Colliery. Layout of the colliery in the 1840s. The T-shaped structure situated in the entrance of the colliery yard is the engine house for the Boulton & Watt pumping engines.

The earliest recorded date for Soundwell Colliery was the delivery of a 26in cylinder and other parts of a Newcomen pump to Samuel Whittuck and Co. in 1750. Six years later the same account book records parts for a second engine which included a 33in x 9ft cylinder.

Information which was given to a Mr Monks in 1914 stated that two Newcomen pumping engines worked on the pumping shaft of Lower Soundwell Colliery. The pump-work in the shaft originally consisted of a series of bucket lifts, and as the extent of the workings expanded, the amount of water increased so in 1756 a second Newcomen pump was installed at the Lower Pit to cope with the additional water. Some of this information was confirmed in an excellent little book by K. H. Rogers who published extracts from Thomas Goldney's account book in 1976, and which provides interesting details of Newcomen engines in the west of England.

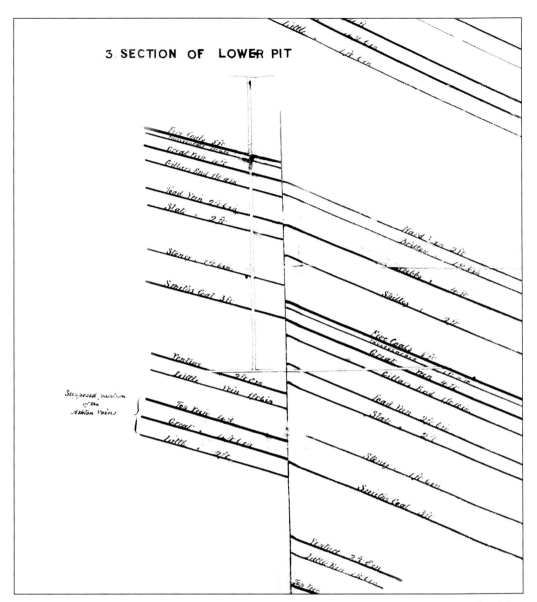

Shaft section of the Lower or pumping pit, showing the workings extending out on either side of the shaft, intersecting a number of seams.

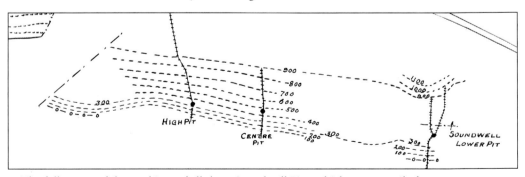

The full extent of the workings of all three Soundwell Pits, which are one mile from east to west.

At a later date a Boulton & Watt engine was installed, but unfortunately the two original pump rods and bucket pumps were attached to the new engine. Curiously there is no mention of any new engine being sold by Boulton and Watt to Soundwell Collieries, so perhaps the new engine could have been second hand. When water broke into the pit in 1853, both sets of bucket pumps were choked by the debris rendering the pump inoperative, thus flooding the pits, which probably had already been worked to exhaustion.

Seams worked at the Soundwell Pits

Five Coals seam 5ft thick normally.
Thurfer seam 1ft 4in to 2ft thick.
Kingswood Great Vein 3ft 6in to 4ft. In some places the seam can be 7ft thick.
Gillers Inn Vein 1ft 10in thick.
Little Toad Vein 2ft 6in thick.
Slate seam 2ft thick.
Stony seam 1ft 6in thick.
Smiths Coal seam 3ft thick.
Hard Venture seam 3ft thick.
Little Vein 1ft 6in thick, probably not worked.

Little is known about the winding engines, but clearly horse gins would have been used for sinking originally. The upper seam (the Five Coals seam) was usually five feet thick, the seam was 450ft deep, which was just about the maximum depth for a large gin powered by several horses. Steam power would have been in use by 1810, particularly when working the Kingswood Great Vein. The Monks' letter tells us that in the 19th century the steam winding engine on the Lower Pit also drove a tail rope down the shaft, pulling coal from a branch 750ft in length driven north to the Gillers Inn seam. In one area the Five Coals and the Gillers Inn veins came together, with coal 7ft thick. Here the roof was exceptionally good, allowing long faces being opened up without having to set timber.

Although an early development, the total area of extraction of the Soundwell Collieries was exceptionally large by 18th century standards. The three pits which were under one manager covered an area which ran from Fairlyn Drive just south west of Rodway Hill to Hillfields Avenue in the west. The mining practice known as horizon mining was in use in the 1780s. At Soundwell this entailed sinking shafts on the anticline and driving headings out to intersect and work a large number of seams. To assist underground working, the steam winding engine on the surface of the Lower Pit also drove an underground haulage system, which consisted of a tail rope pulling coal from a branch 750ft in length, driven northwards to the Gillers Inn seam. This haulage system worked down through the winding shaft into the workings. In fact the Soundwell Collieries were technically very advanced and were one hundred years ahead of other Bristol collieries.

In 1845 there was a serious accident in the Upper Pit in Gladstone Street, when four men were killed by falling down the shaft. Thomas Bird, William Bassett, Ben Wiltshire and John Porter were riding in a hudge when the rope broke throwing them down the shaft. It was in the western part of the take under Hillfields Avenue, close to the old workings of the Lodge Hill Pits, that the barrier between the pits was breached and water broke in, flooding the workings. The water ran downhill to the Lower Pit which was the lowest point of the workings and choked the pumps. The colliery was flooded and ceased work sometime in 1853.

Sadly there were three fatal accidents in the last two years of working. One was the haycock or haystack boiler bursting and killing the banksman, a Thomas Waller, on the 12 December 1851. On 24 March 1852, George Morgan a collier, aged 18, fell out of the hudge and fell down the shaft and was instantly killed. On 29 December 1852, a collier, John Green aged 56, was killed when the winding rope broke.

The New Cheltenham Pit

This small but interesting pit, situated near the bottom of Spring Hill, only worked for a short time. The Coal Commission report of 1871 states that the Stony seam two ft thick was worked at a depth of 156 ft. The Smith's coal is also mentioned as being 22 ft below the Stony seam, but it is not clear whether this one was worked. The last date on the abandonment plan was 29 December 1873, but the pit is listed in the Inspectors List of Mines in 1874-1877 under Chubb & Waring, but in 1878 the pit is shown as not working. Presumably Chubb & Waring were hoping to find a buyer for a pit, with little or no prospects. The chances of extending the area of workings for this pit were poor, as old early 18th century workings were present to the north, east and south. New Cheltenham Pit was also at risk of flooding from the old workings situated to the north, as they were 80 ft higher than those of the New Cheltenham Pit.

An Area of Early Mining

About 1,200ft to the north-west of New Cheltenham Pit is an area of early mining which, on an old map dated 1750, is shown to be in Sir John Newton's Liberty. The area was known as Cockshot Hill on maps of 1883 and this old Kingswood name still survives today, also known as Lees Hill. In 1684 a number of coal pits were at work or abandoned. One was owned by a Jonathan Tucker which is known to have been left open in 1684. The same owner had four pits at work at Garrotts Mead (several hundred feet to the west) with a water level. The author well remembers playing on small spoil heaps in a field between the Shant (a public house) and Lansdown Road.

Years later site investigations revealed a number of bell pits in the grounds of St Stephen's School. These small coal pits were found to be 50ft deep, one was even 90ft deep, and flowing water was located in one bore hole at a depth of 90ft. Were these pits the coal works of Jonathan Tucker? The water flowing at 90ft is possibly the water level. They were certainly the features over which I played, and appear never to have been recorded although there were clearly shafts visible in the 1950s. These workings were in the Kingswood Great Vein, Thurfer and Five Coals seams. Just a short distance to the north-east was found the largest bell pit the author has ever seen. It was 18ft in diameter, and would have been worked in the 1670-80 period. All of these old workings, which must be waterlogged, are situated north and 80ft above the old workings of New Cheltenham Pit.

Potters Wood and Jays Pits

Potters Wood and Jays Pits are situated in an area that was worked from about 1670. These pits were worked by a number of different owners from the late 18th century until the final working by J.J. Whittuck in 1869. Many of these early pits are now just a name; today even some of their positions are now lost and few people now living at Potters Wood would have heard of Nan Haine's pit, Tom Jay's Pit, Ned Fryer's, Milk Street Pit, and Tun Pit and many others. They were small concerns owned by individuals who often gave their names to the pits they sank and worked. By 1750 some of the coal pits at Potters Wood were dewatered by Potters Level which certainly drained the two Potters Works. Potters Wood Pits are situated 1,500ft to the south of the main Two Mile Hill to Kingswood Road. The shaft of Potters Pit was originally 293ft deep and Jays shaft was 277ft in depth before the 1860s.

Both were equipped with small single cylinder horizontal winding engines with egg-ended boilers. In the later period these pits were raising 60 to 70 tons of coal per day. Potters Pit had no guides and worked with a free swinging bucket, but Jays Pit had a single cage running in wooden guides. Both pits were linked underground at the 100 yard level and joined the adit

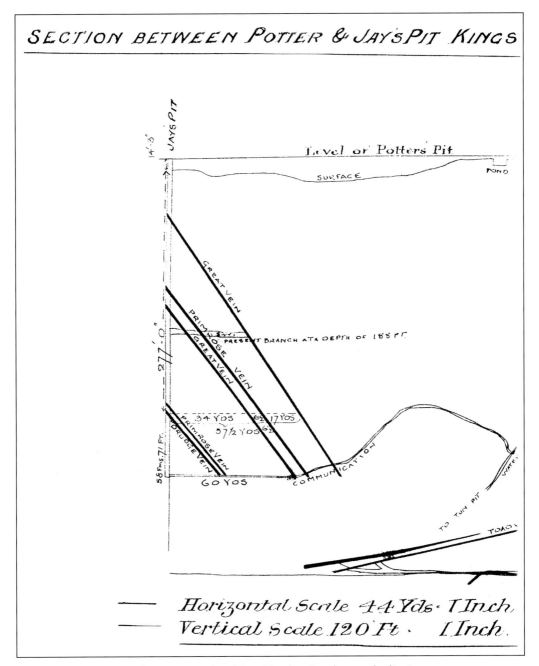

A section of Potters Wood and Jays Pits showing the steeply dipping seams.

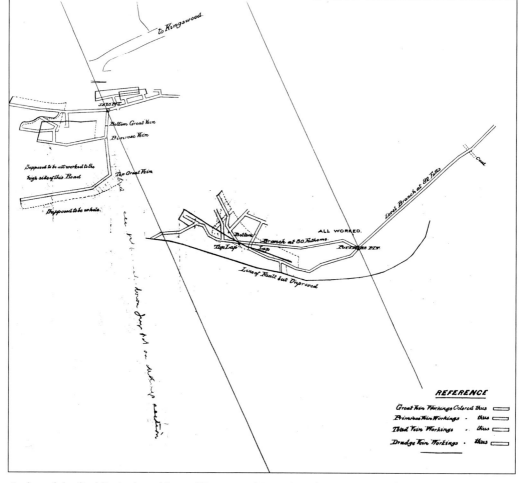

A plan of the final limited workings of Potters and Jays Pits, after 100 years of working in faulted and steep seams which were difficult to work. The closure of the pits would have been no surprise.

level about 30-40ft west of Potters Pit. Early records, destroyed in the late 1950s or early 1960s by a local firm showed that from Potters, and other shafts to the north and east, the Kingswood Great Vein was totally exhausted to the depth of the free drainage. The Potters Wood Pit was deepened from adit level to 480ft, and a branch was driven south intersecting the Kingswood Great Vein at this depth. These deep workings were unfortunately lost due to water, which exceeded the capacity of the water bucket and the engine winding it with third motion. By this time it was an extremely antiquated engine.

Clearly the lessees had no capital to provide more up-to-date machinery. Coal was last raised from the 100 yard level in 1866. Jays Pit was also deepened from the 100 yard level to 120 yards and a branch driven in a southerly direction. It crossed a number of seams:

The Drudge
The Gillers Inn seam
The Kingswood Great Vein

This branch was driven further in the hope of finding another lap of the Great Vein, but the pit was closed before any further coal was cut. At Potters Wood the Drudge was a good clean seam from 16in to 18in thick. It had a good roof, but the workings only extended a short distance on either side of the branch. The Gillers Inn seam was 4ft thick, with good roof and in three beds with beds of shale partings. The Kingswood Great Vein was a clean hard coal 7ft thick with a good roof, and was worked at the 100yd level, and down to the lower workings. At these lower workings, the vein was opened from the branch level to the west where it pinched to 2ft. It also became thin in the east of the branch. Between this east and west thin coal, the Great Vein was from 6 to 11ft thick in one clean bed, with the roof of the workings standing on Billets and Traps. Deep side gugs were driven in which the vein maintained its thickness, but yielded a little water for which there were no pumping appliances available.

There was a plan in 1868 to open out the pits for deeper development, which entailed the dewatering of Potters Wood Pit, opening the Great Vein through the 160 yard level south branch, and communicating with Jays Pit by an incline on the Great Vein up through the workings at the 120 yard level. It was hoped that the removal of water from Potters Pit would drain the deep side gugs at Jays Pit and allow communication to be made from this side also. In pursuing this plan, all coal was raised at Jays Pit; the water at Potters Pit was raised by a water bucket, which was discharged into the adit. After several months of winding, the water was lowered to about 160 yards, when suddenly the adit level flowed backwards and filled up the pits for some distance above the free drainage and drowned all of the workings.

This influx of water from the adit level was found to be due to operations at the Avonside Level, where the lessor had permitted some persons to open up the Rag Vein in the adit level between the upcast at Hanham Collieries and the Hanham Pit. They broke down the arched roof of the adit level for over 40 yards, and formed a road in the roof and dammed the water, except what could pass through a small pipe they had placed in the floor. The water was found issuing from the pipe with great force, indicating a considerable head of pressure. This obstruction in the adit level was removed, when the water at Potters Wood and Jays Pits at once fell to the 100 yard level, and the water in the adit resumed its natural flow. The pits were then unwatered, Jays to the 120 yard level and Potters Wood a little lower completing the deep development.

An accident at Potters Wood Pit occurred which stopped all effort to unwater the pit with a bucket. Neither steam power nor pumps were available. It was also found that the underground workings and the communications between pits had been greatly damaged by the flooding, and could only be reopened at considerable cost. Considering the estimates of the expenditure that had been made, it was found that the amount of capital required to carry out the deep developments exceeded what the lessors and other interested parties were willing to provide. So it was decided to close the pits, which was done in August 1869. Later the lessor disposed of the small engines and boilers, and the pits were abandoned. In 1980 the site of Potters Wood Pit had no structure or anything of the pit visible except for the filled shaft, which was capped.

California Collieries

California Colliery was an old shaft originally sunk to the thin Francombe seam which is only 10in thick and was 240ft in depth when abandoned. It was reopened around 1875 by Mr Abraham Fussell and became the Oldland Colliery Co. working the four seams of the Upper series, the Millgrit, Rag, Buff and the Parrot veins. The old shaft became the downcast and was deepened to 522ft, and a shaft known as Peg House and situated 1,300ft to the north became the upcast shaft with a depth of 474ft. A modern twin cylinder winding engine was installed, with cylinders 16in diameter, a 3ft stroke, and a drum of 9ft diameter. Five Lancashire boilers provided steam, two of which were made at Phipps' works at New Cheltenham. The pumping engine was also made by the local foundry, Gregory of Kingswood Hill. It was a direct-acting condensing engine with a cylinder 60in in diameter, a 9ft stroke. There was also an air pump 32in diameter, 4ft 6in stroke. The wrought iron beam was 15ft in length, and there were 3 Cornish valves. There was a wooden engine house with cast iron windows.

More Gregory plant was installed in the compressor house, in the form of a steam pump with 12in cylinders. The surface compressor house contained a pair of horizontal air-compressing engines with steam cylinders 21in diameter, air cylinders 21in diameter and a 2ft stroke with a massive flywheel 14ft in diameter. This equipment provided compressed air for the underground haulage engines and to the Cowhorn Hill and Brook Pits. The headgear was made from pitch pine 28ft high with two sheaves each 9ft in diameter. The No. 2 or upcast shaft also had a wooden headgear 20ft high with a 6ft diameter sheave. The winding engine was probably portable made by Garrett & Son of Leicester with two 10in cylinders, 12in stroke with a drum 6ft in diameter geared 1 to 6.

There was no rail connection to the pit, so an incline was constructed connecting with the Avon and Gloucestershire Railway, which carried the coal down to the Avon. A land sales wharf was also opened at Willsbridge. On 21 May 1900 a workman was injured by an overwind when he ascended the upcast shaft, this being the result of gross carelessness on the part of the engine man. The inspector's report stated that the drum on the engine was only 6ft in diameter and was geared 3 to 1. The head frame was 18ft high.

The colliery continued working until 1904. By then most accessible reserves must have been exhausted, but in March an inrush of water from old workings flooded the pit in a very short time. In the early 1960s I met a Mr. Burge who was then in his late 80s, living near the site of Camerton Colliery. He told me that as a young collier he had worked in California Colliery; he always hung his coat (containing his week's wages) on a pit prop near his working place, when the water burst into the workings, and the water rose so fast he did not have time to recover his coat and wages. Other men told tales of losing their watches and other personal effects in the flood. The sale particulars for the 16 February 1905 inform us that the collieries worked until March 1904, when the colliery was voluntarily wound up. A total of 300 miners were employed and as much as 1600 to 1700 tons of coal per week would have been raised. There was no mention of the flooding which had closed the colliery.

Opposite top: California Colliery workings in 1904 when abandoned due to flooding. Most of these workings were in the Millgrit and Parrot Veins, with limited workings in the Buff Vein.

Opposite bottom: California Colliery about 1900. It is likely that the small tin shed with cast iron windows held the Gregory Pumping engine.

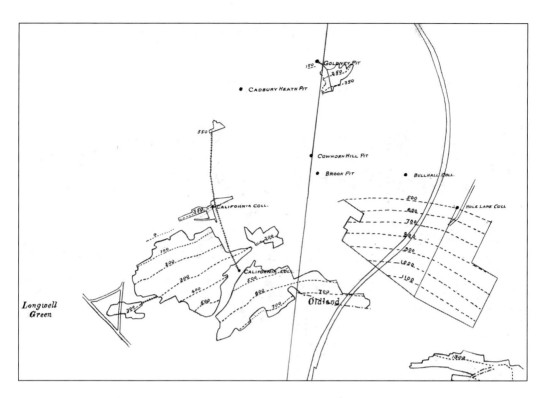

Hole Lane and Bull Hall Collieries

Bull Hall Colliery is most likely the earlier of the two collieries as it was sunk nearer to the shallow coal and outcrops. Hole Lane Colliery was certainly at work in 1820 as was Bull Hall Colliery. Hole Lane appears on an early map as Haul Lane Coal works and may have been sunk as early as 1810-20. The owner of this small pit in 1841 was a Robert Jefferis who also worked a second pit known as Hole Lane at Harveys Lye to the east of Westoncourt Farm and north of Redfield Hill. He also owned Cowhorn Hill Colliery. The three pits employed 150 hands in 1841, forty of whom were under 13 years of age, the youngest being 7 years of age, having started work the year before. On a plan of 1830 Haul Lane, as it was then spelt, only had a horse gin for winding, no steam engine can be seen, but at this date the colliery and Bull Hall Pit already had tramway connections. The depth of the shaft was 661ft, and Bull Hall was also the same depth. The shaft section is as follows:

Total depth to seam

		Yds	Ft	in
The Millgrit seam	3ft	101	0	0
Rag seam	1ft 6in	116	1	6
Devils seam	- -	186	0	0
Buff seam, variable	4ft 0in	193	1	0
Parrot seam	1ft 9in	220	1	9

Both Hole Lane and Bull Hall pits were sunk to the four seams of the Upper coal series, the Millgrit, Rag, Buff and Parrot veins. The last seam was in much demand by the Brass Industry as being sulphur-free and was sold for 15 shillings a ton at Keynsham in 1870. It is not known when steam power replaced the horse gin, but once the pit was steam driven the same engine (a rotative beam) would have worked until the pit ceased working. The abandonment plans show the two pits as having been closed in 1873. However they are shown in the Inspectors' List of Mines as Bull Hall Colliery and Hole Lane Colliery under the ownership of Jefferis in 1875. In 1876 both pits were owned by Fussell & Co. In 1879 and 1880 Hole Lane Colliery disappears from the Inspectors' List of Mines. Today the blacksmith's shop still stands at Hole Lane and one or two small buildings remain as buildings at the bottom of gardens. The shaft capping was breached by a builder in the late 1960s and the shaft was filled and capped by the National Coal Board.

The only standing remains of Hole Lane Colliery today. This building housed the blacksmith's shop.

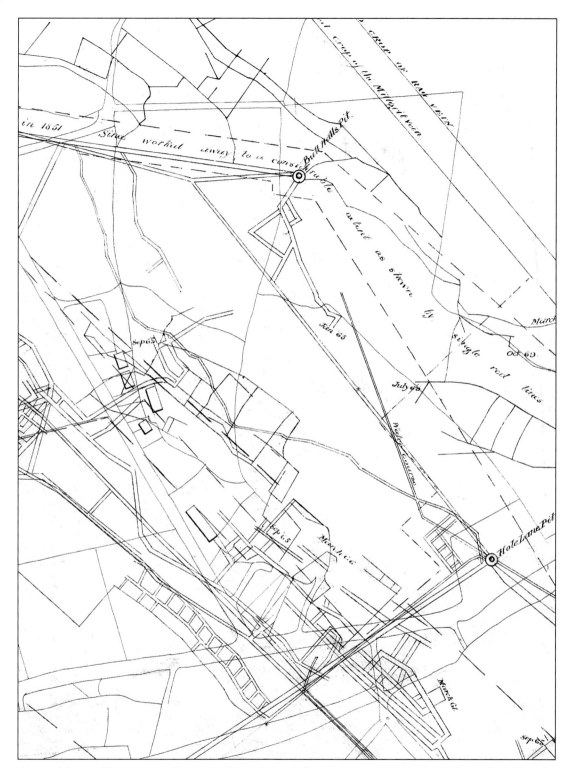

Some of the workings near to the shaft of Hole Lane Colliery with a water course which ran for half a mile to the west to the Cowhorn Hill pumping engine. These workings are in the Parrot and Millgrit veins.

Cowhorn Hill, Brook and Buff Pits

Cowhorn Hill and Brook Pit are old ones which are known to have been at work in the early 1800s and may have been of late 18th century sinking. Cowhorn Hill had a Boulton and Watt pumping engine and a Boulton and Watt double-acting winding engine of 11 hp. By 1841 Cowhorn Hill Pit was working the Buff vein 6ft thick, and a seam 2ft 6in thick. A steam pump of 36 hp worked 16 hours out of 24. A second pump at the deep pit was 64 hp. The shafts in 1841 were 108 fathoms, 60 fathoms and 34 fathoms which are 648ft, 360ft and 204ft respectively. The pit was equipped with cast iron plateway and wheeled tubs containing about 2 cwt of coal were pulled along this, although small boys would bring out coal from the stalls in carriages holding a bushel and a half each, sliding on wooden ladders up an ascent of 2ft in 12. The ladder was formed of plank laid on the floor about 16 inches wide with cross-bars, by which the boys ascended on their hands and feet, hauling the tubs after them by girdles of rope to which a chain was affixed passing between their legs. If under 50 yards, 104 tubs was considered a fair day's work, for which the pay was 20d, whether one boy or two. A strong boy could do this work by himself, but clearly some boys were not strong enough. The report also states that from 50 yards to 150 yards the same is paid.

The precise position of the Buff pit is not clear: it may be that the Buff pit is another name for one of the shafts at Cowhorn Hill Pit. The abandonment plans only show one shaft at Cowhorn Hill and one shaft at Brook Pit. An old shaft at Cadbury Heath which is shown on the tithe map of c.1840 as having a horse gin, and is 1,100 yards from the California Colliery. This was reopened by the Oldland Coal Company and a new engine house was erected with a pair of horizontal cylinders 20in diameter, 4ft stroke and a drum 7ft 3in diameter. This engine wound from a depth of 339 feet. The headframe was of pitch pine. This pit was linked to California Colliery and California was linked to Cowhorn Hill which had in turn linked to Hole Lane and Bull Hall Collieries.

All other shafts of the Oldland Coal Company were abandoned on the flooding of California Colliery in 1904. One curious mine entry is the Fig Coal Drift which at the time of closure contained 140 feet of steel rails at about 18 pounds to a yard, and a timber bridge over the brook containing 1 ton of timber. This drift is stated to be near the Brook Pit. It may have been driven just before the closure of the concern, or it could be the drift shown on the abandonment plan as a water level on the Fig Seam just south of Brook Pit.

The California shaft and Cowhorn Hill shaft were acquired by Bristol Water Works who installed steam pumping engines and drew water from the shafts until the late 1960s. The buildings at Cowhorn Hill were built by the Water Works and are not original colliery buildings.

Goldney Pit, Cadbury Heath

This Goldney Pit is not to be confused with the Goldney Pit to the north at Warmley. The sinking date for this small pit is not known but was probably around 1800-1820. It is shown on the tithe map of 1840-41 as having a substantial number of buildings which would have included a steam engine house. When the estate now known as Roy King Gardens was under development in the 1980s, foundations of the bob wall of a beam engine were uncovered. Unfortunately these remains were not seen by anyone with a knowledge of early steam engines, so it is now impossible to say whether the remains were of a winding or pumping engine.

The abandonment plan of Goldney Pit shows that the shaft was sunk around 1820 to a depth of 360ft and was reopened and deepened in 1906 to a depth of 567ft. The shaft was 9ft by 5ft and was divided by a stone partition down to a depth of 264ft, the remainder of the shaft being divided by a wooden partition. Curiously the upper section of shaft with a stone partition was only walled on the western side of the shaft. The newer deepened section was circular and 10ft

6in diameter. The reopened pit worked the New Smith's seam which was 2ft 2in thick and the deepest point of the workings was 177 yards below the top of the shaft. The colliery ceased work in August 1909 and the colliery plans show that old workings were encountered just north of the shaft at a depth of 240 feet, which would have been the original workings of c.1820. To the north of Roy King Gardens several oval shafts were uncovered which would have dated from about 1810-30, and to the south of the housing site several bell pits were uncovered on the outcrop of the Cuckoo Vein. These bell pits would have been worked between 1680-1720. The colliery ceased work in August 1909.

The Grimsbury Pits

The pits consisted of five shafts, some of which were sunk in the mid-18th century. One shaft which slumped some years ago revealed a square shape with rounded corners which is typical of 1750-60 date. One of the other shafts was said to have been shaped like a bishop's mitre. One other shaft of this shape is that of Shoscombe Colliery in Somerset. The only depth is for Bayntuns shaft which is recorded as 480 feet to the Great vein which was found to be 3ft 6in thick. The Grimsbury Pits also had a set of coke ovens along side a striking engine house which still stood in 1950, and was then unfortunately demolished and used to fill the winding shaft by the National Coal Board some time afterwards. Although no plans exist, the only section of the Grimsbury Pits shows that the workings cover a large area some 3,000ft from north to south. When the ring road was under construction a cast-iron wind-bore from a Newcomen-engined pump was found which might have come from one of the Grimsbury Pits. This is the only wind-bore of a Newcomen engine ever to come to light. The casting has cast on it the maker's name – Jones & Co. of Bristol and the date on it was 4 March 1777.

A nearby pit was the Tennis Court Pit on the southern end of Tennis Court Road which was sunk by Monks on a fault, and it is said that no seams were met with. Had Monks driven eastwards away from the fault, towards the Kingswood Grammar School he would have found workable seams below the many bell pits sunk in the early 18th century.

One of the shafts of the Grimsbury Pits in which the fill had slumped. The shape is typical of the 1760 period.

The excavation of the foundations of a ventilation furnace on one of the Grimsbury Pits, by the Bristol Industrial Archaeological Society in the 1980s. Clay pipes, found on the site dated from the 1840s, were probably dropped during the demolition of the furnace. The site only contained an air shaft and the furnace. The shaft, although not examined, appeared to be distorted and may have collapsed.

The only known windbore of a Newcomen Engine ever found, this windbore was found when the ring road was under construction, and may have been installed on one of the Grimsbury Pits as it was found 1200ft from them.

This windbore was made in Bristol by Jones and Company at Cheese Lane [40 St Philips Plain]. The date of 4/3/1777 was chiselled into the pipe. The windbore was the lowest part of the pipe which sat in the sump, the bucket pump was situated above and was built into a separate casting.

Crown Colliery, Warmley

This interesting old colliery was sunk sometime around 1820 to the south-east of Siston Common Colliery and north-east of the Grimsbury Pits. The colliery was probably sunk when the Grimsbury Pits were nearing the end of their life and the Siston Common Pits were encountering problems from disturbed and faulted seams. The colliery was comprised of four shafts :-

The No. 1 shaft, depth 492 feet, shaft diameter 7ft, now with a concrete cover.
The No. 2 shaft, depth not known, sealed with concrete cover.
No. 3 shaft, depth said to be 480 feet, on section dated 1890 which shows the depth as 540ft.
No. 4 shaft, depth 390 feet sunk through faulty ground on south side of shaft. It appears to have worked to the rise only, due to faulty and uncertain ground.

    Two shafts were north and south of the main London Road and much earlier than the two other remaining shafts. The shaft to the north of the main road was the engine or pumping shaft which had a 50 hp beam pump, probably of the Watt type. The coaling shaft was the No. 3 pit: with a 22 hp engine which used a plaited rope; a running stage was on the shaft.
    In 1841 the proprietors were Davidson and Waters who employed 60 hands. Of these 11 were under 13, the youngest was just 9 years old. The colliery was purchased around 1866 by G Goldney who worked the pit for some years. It was connected to the Avon and Gloucester Tramway but by 1883 this spur into Crown Colliery was long disused. Towards the end of the collieries' life the concern had a number of short-term leases. One concern tried to work the coal deposits to the north-east of the take at Webbs Heath, driving a drift mine back down to the south-west and avoiding the area of 18th century mining around the crossroads at Webbs Heath. Unfortunately this drift was driven at the wrong angle, and after several hundred yards the project was abandoned in January 1901. The real reason for the abandonment was a large fault, known as the Church Fault which ran north-east from St Barnabas Church. A large area north of this fault is shown as faulty and barren. It was probably this fault and the large amount of barren ground, with additional problems caused by Midland Railway not allowing the colliery to work under the railway line, that caused work here to cease.

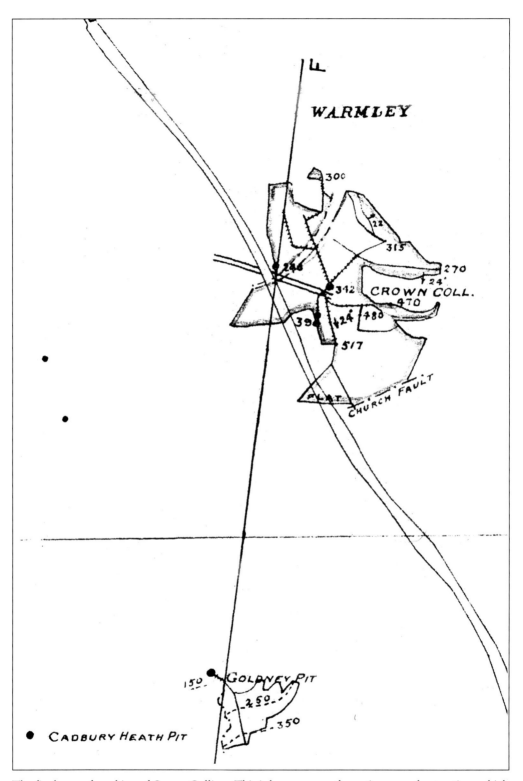

The final area of working of Crown Colliery. This is by no means the entire area of extraction, which would have been 50 per cent larger, as the pit worked for some 80 years. The small take of Goldney Pit, Cadbury Heath, is situated to the south.

The engine house and chimney of Crown Colliery, built around 1880 for the last phase of working, and now sadly demolished.

Crown Colliery workshops, now the only surviving structure of Crown Colliery. The shaft is filled and is in this yard.

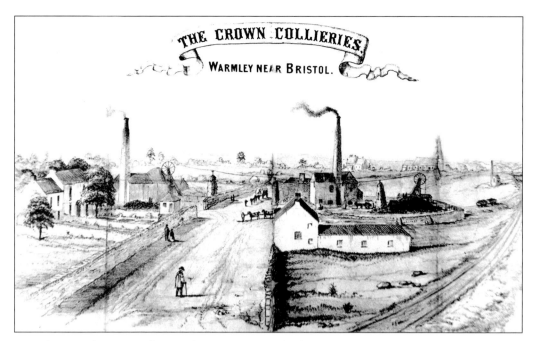

An early print of Crown Colliery in the 1850-60 period. The pumping engine is on the north side of the main London Road with the main coaling shaft on the south of the road. The pit in the distance is the early Goldney Pit.

The engine house for a drift mine on Webbs Heath, the last phase of Crown Colliery which was abandoned by the 17th January 1901. Taken in the early 1960s, little now remains except for the chimney.

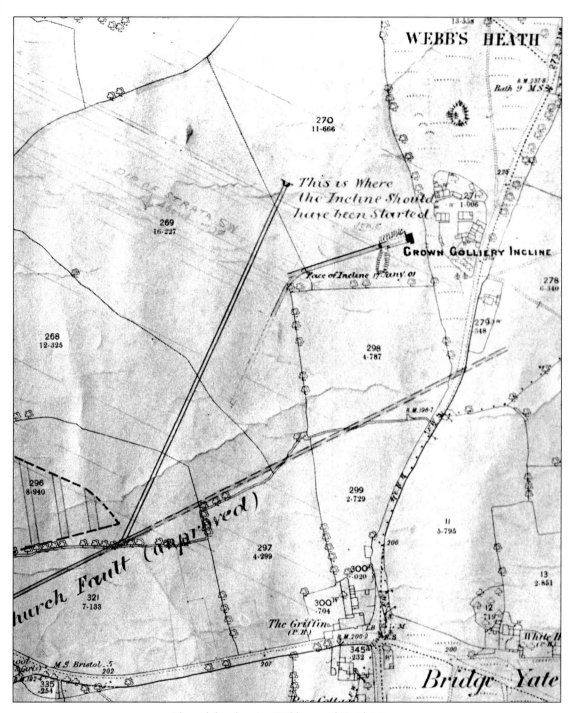

Plan of the abortive Webbs Heath Drift Mine.

Goldney Pit

This pit is situated just south of the Crown Colliery at Warmley, and should not be confused with the pit of the same name at Cadbury Heath. The sinking date for this small concern is not known, but in 1830 the pit had a horse gin for winding and a steam pump. An engraving of Crown Colliery, not dated but probably from 1860, shows an apparently disused engine house nearby, similar to the one at Grimsbury Pit. Nothing is known about this pit except that it was sunk over a coal seam outcrop 500ft south of the South Pit at Crown Colliery, and was 300ft north of the Church Fault. It must have had a short life working the seams below those of Crown Colliery.

The remains of the brick foundations of the 1900 period winding engine house of the Goldney Pit at Cadbury Heath, taken in 1965.

Syston Common Colliery

The sinking date for this colliery is not known but is likely to have been around 1790-1800. As expected, the colliery originally consisted of one shaft with a horse gin. By 1841 the horse gin was still in use, but a building which appears to house a steam engine stands opposite the horse gin. A prospectus of 1889 tells us that the colliery then had four shafts two of which were sunk to a depth of 95 fathoms (570ft) at a distance of 25 yards apart, and have cut through several seams of good house coal. From various estimates it appears that there were no less than 6,500,000 tons of workable coal, after making allowances for waste. It was also estimated that 3,000 tons of coal per week could have been raised for some 45 years. A quick look at the geological map will show that these claims are far from the truth. The whole area is shown to be highly faulted with numerous thrust faults. Some 700 feet to the south-east is the most highly faulted area to be found in Kingswood. Here the coal measures are not only vertical, in some places in the old clay pit of Haskins Pottery they were totally inverted and crushed and unworkable. The plans of Crown Colliery at Warmley show that the faulting increased to the north with areas of barren ground. It was surprising that respectable mining engineers allowed their names to be incorporated in the prospectus. For Thomas Brown to say that the collieries have only been slightly worked is surprising as it was known that the colliery had been working for almost 100 years. For an exhausted colliery the plant and equipment was surprising lavish. The winding shaft was 9ft in diameter, walled throughout and fitted with girder guides which were made of the best pitch pine, and two separate cages. The winding engine was claimed to be a first class double horizontal, with 26 in cylinders, 4ft 6 in stroke, a 12 feet drum for round rope and was capable of raising from 700 to 800 tons of coal a day. There was a powerful compound engine for pumping with two cylinders, 19 in, in diameter, new gear wheels 3 to 1, pipes and condenser all in working order. The chimney stack was 100 feet high and built of bricks, and connected to three egg-ended tubular boilers. A report by Messrs Needham of Newport gave a glowing view of the enterprise, but does say that they had no record of the exact quantity of coal worked. Needless to say the colliery never reopened. The plant was sold off, the buildings were demolished and the shafts filled. Today the site is as it was, with a small spoil heap, remains of heapstead and walls of the pumping engine house, and a capped shaft along side the abandoned Midland Railway's Bristol to Bath line.

The engine house at Syston Common for a Newcomen pumping engine, built of copper slag blocks. The engine was certainly at work as early as 1777.

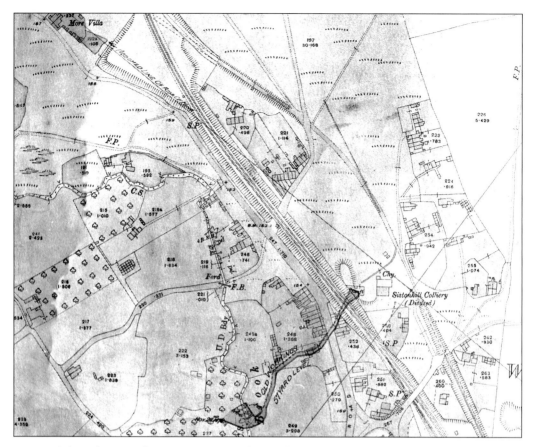

A first edition Ordnance Survey map of 1883 showing the site of Syston Hill Colliery.

The remains of Syston Hill Colliery in the early 1960s showing the former position of the pumping engine house and shaft just east of the railway line from Mangotsfield to Bath.

The Harry Stoke Drift Mine

The story of Harry Stoke Mine commences with the closure of Speedwell Colliery in 1936. The accessible reserves here were over a mile from the pit bottom. The main reserves were in the Kingswood Great Vein which had become increasing expensive to work due to the increasing cost of underground haulage with the problems of having to work with a large number of haulage engines on separate inclines. The reason for the closure of Speedwell and Deep Pits was given on the abandonment certificate by Arthur Savage the colliery surveyor, as been unprofitable and lack of capital, or plainly worked out. After Frog Lane Colliery at Coalpit Heath had closed in 1949, this was the last colliery at work in the Bristol portion of the coalfield. So in the 1950s the National Coal Board drew up plans to open up the northern section of the coalfield, one mile to the north of the abandoned workings of Speedwell Colliery. This post-nationalisation venture was situated in an area which was unique, in that the surface strata is of the Triassic period which is newer than the Carboniferous series to which the coal measures belong. The intervening Permian Series is absent. The Triassic beds rise to the east, and the coal measures beneath dip unconformably to the east. The mine consisted of a pair of drifts which passed through the Triassic beds at a gradient of about 1 in 3.3 to enter the coal measures along their natural dip. The original plan was that flank faces be driven and worked from the new dip development in both directions for a distance of 1,500 feet. The new drifts were to be driven down 3,000 feet along the crop and from these to extract 1,500 feet on either side, and then yet another drift mine for the next adjacent 3,000 feet panel.

As the drift mines reached the deeper measures, shafts were to be sunk beyond to extend the extraction further. It was proposed to sink two shafts at Downend, and it was hoped that the Harry Stoke Mine would have been the beginning of a major scheme of development for the industry. Three seams were accessible from the drift: the Five Coals seam, the Kingswood Great Vein and the Gillers Inn seam. The Five Coals and the Gillers Inn seams were about 100 feet apart with the Great Vein intervening. The upper seam, the Five Coals, was the most extensively worked, and was normally the best to work. Panels were set both right and left of the dips, but each stopped eventually due to disturbance, sometimes in the form of barren ground and sometimes because of thin coal. In 1959 the pit made a profit of 1s 6d per ton amounting to £9,200. This was achieved by four good months, January, February, March and April, with a cumulative profit of £21,400. A sum of £12,000 was later dissipated by increasing losses month by month.

A limited amount of work was undertaken in the Kingswood Great Vein, and then attention was switched in 1961 to the Gillers Inn seam. A 480ft plough face was prepared in the Gillers Inn seam in 1961, but due to a friable roof and fluctuations of 12in to 15in in the seam section, the output was low and the quality very poor. The Gillers Inn seam had to be worked by hand after attempts to power load failed. These faces encountered disturbances, and the seven faces that had worked the seam did not achieve the advances that the Five Coals faces had done. The M5 face met disturbance in October 1962 although the seam section was 4ft 9in. The M7 face was the main production face and was 381 feet long, but 180 feet from the right hand gate, barren ground appeared, affecting a length of 111 feet. It was the presence of this barren ground that precipitated the proposed closure. Another face the H9, was 270 feet long, and had section of 4ft 8in, but a 5ft fault crossed the face. Although the fault was manageable, production was limited due to the inability of the gate belt conveyor to carry coal up the gradient.

Papers dating from 1962 show that the mine had produced 470,000 tons of coal with a total loss of £818,000 or 35s per ton overall. In 1962 the mine made a loss of 14s 6d per ton and a total loss of £50,000. National Union of Miners' documents suggest that insufficient effort had been made to penetrate the disturbances and that the emphasis had been to maintain an output of 300 tons at all cost, with the result that new faces were opened without sufficient effort having been made to re-establish old ones. The Board replied by saying that it did not

The timber yard at Harry Stoke in May 1961.

Harry Stoke Mine in May 1961, with C.A Francis first aid attendant; Alb Gill, first aid officer for Somerset & Bristol Group, and Lewis Watts who was an overman at Pensford Colliery.

Salvaged or scrapped rings at Harry Stoke mine, 1961.

The northern drift at Harry Stoke Mine.

The elevated tub line from the No. 1 drift to the screens and washery, taken in 1960.

appear that continuing successful operation could be anticipated within the horizons as yet worked, and if the mine were to become economic it must extend further into the coal measures. NCB engineers felt, first, that there was no indication that the deeper measures would be better, and secondly, the strata dipped over at the extremity of the drift, so that conveying working would become more difficult.

It is surprising that the original engineers, when planning the Harry Stoke Mine, did not take into account the geology and condition at Speedwell Colliery some 20 years previously. An appraisal of these workings would have raised doubts on the viability of the Harry Stoke project, particularly as Arthur Savage had been the senior surveyor at Speedwell Pit at the closure. A document prepared by the NUM stated that:

'The question arises, is a loss of £318,000 sufficient grounds to justify not the abandonment of a mine but of a coalfield, when facilities are available to probe deeper and manpower is available and capable of such a high rate of production under conditions that are not easy. There are no problems at the mine other than the absence of regular seam conditions. As yet the main drift is down 1,000 yards or to a depth of about 300 yards below the surface. In common with other Somerset mines the proceeds are low and consequently high output rates are necessary. Power loading has helped to achieve these high rates in the remainder of the coalfield and obviously the ability to power load at Harry Stoke would quickly change its economic position. It is not practicable to power load at the present horizons but one cannot rule out the possibility if workings were advanced further from the crop'.

This statement from the NUM is interesting as only six years later the shortage of men led to the closure of New Rock Colliery and Norton Hill Colliery, the biggest pit in Somerset. In the 1960s Harry Stoke was remote from the rest of the coalfield, and the men on the surface were mostly disabled. In those days the NCB had a policy of looking after men who had had accidents, and they were almost always found jobs on the surface of mines. If and when Harry Stoke closed all of these men would be redundant. The underground men would travel by coach to other pits like Norton Hill, Kilmersdon and Writhington, but this would take a travelling time of 1 hour 55 minutes before reaching the colliery. Many would have to travel by bus to the pick-up points, whilst others could not reach the coaches. The NCB realised that a large number of the men would not find employment in the industry. They accepted that the social consequences were likely to be greater than any arising out of any previous planned closure in the South Western Division (South Wales & Somerset).

The saleable output in 1960 was 65,927 tons, with a loss of £48,200; the last quarter having made a loss of £21,000. The reserves at June 1961, were said to be 200,000 tons, and the Board said that to extract the coal, extensive exploration and drivage would have been needed. Bore hole data held by the Coal Board indicated that disturbed ground continued and the seams might well deteriorate further. The greater depth would entail the sinking of a ventilation shaft and increased transport would be required. The Board also felt it would not be justified in risking the throwing of so much more good money after bad. A letter from the Chairman of the NCB stated that the friable roof conditions made mechanisation almost impossible and the high ash content, which was 30 per cent, was no longer acceptable to the Central Electrical Generating Board for their modern design of power station furnaces, so the possibility of working these two seams offered no hope of producing a more marketable product.

From the production and marketing viewpoint therefore the Board came to the conclusion that they must accept the Divisional Board's recommendation that the colliery must close. It employed just under 190 men of whom 40 were employed on salvage work for about 3 months, and 40 were offered immediate employment at other pits in Somerset. The rest of the men who were willing to move were offered employment elsewhere in the South Western Division, with the full benefits of the Board transfer scheme. The men who were over 60 received an immediate pension under the Mineworkers' Pension Scheme, in addition to redundancy compensation.

This closure on the 14th June 1963 brought to an end the northern portion of the Bristol Coalfield, and the southern or Somerset section would follow in 1973. This was not quite the end of the story: in 1979-80 I was underground at Daw Mill Colliery in the Midlands when I was approached on the face by an ex-Somerset miner who recognised my Bristol accent. He proceeded to tell me that there were several other Somerset miners still working there underground.

Production 1954-63	Manpower	Saleable Output, Tons
1954	107	4,430
1955	160	12,996
1956	178	42,507
1957	276	46,507
1958	299	71,125
1959	not known	101,000
1960	not known	not known
1961	21	47,025
1962	206	65,900
1963	174	16,076

The final closure occurred on the 14 June 1963.

 # The Mangotsfield Collieries

Deeds and leases exist for the working of coal in Mangotsfield in the 16th and 17th centuries. Unfortunately the location of some of the late 17th century coal pits is now unknown. They are listed as:

> "Mangotsfield Green 1 pit,
> in bottom of said green 1 pit,
> and near Will Ploomers House 1 pit,
> and 1 pit near to Thomas Hills House."

There was certainly a large amount of mining in Mangotsfield in the 18th century, which has gone largely unrecorded except for a few deeds and leases, for which the locations are not clear.

The first known date of mining on Mangotsfield Common was in 1875 when an entry in the Mine Inspectors List of Mines gives an S. Phipps as being owner and manager of Mangotsfield Common Colliery. The workings of the mine run by Mr Phipps were not extensive and after some time the mine filled with water, resulting in the pit standing idle. In the early 1900s the Bristol Wallsend Collieries Limited reopened the colliery with a T.H. North as Manager. The only seam worked was the Mangotsfield Great Vein which ranged in section from 3ft to 2ft at a depth of 100 feet, but was of very poor quality. It was finally abandoned in November 1907. A letter in 1910 from Mr North to the Mine Inspector Mr Joseph Martin is most interesting, as he refers to this most disastrous property. It becomes clear that the previous owners sank the pumping shaft deeper than it need have been; the Mangotsfield seams were only 30 feet below the surface although the pumping shaft was sunk to a depth of 100 feet. The original company did not have a clear understanding of the geology at Mangotsfield, which is affected by series faults, between which the seams are increasingly displaced to the west. Unfortunately the Mangotsfield area and Mangotsfield Colliery is affected by one of these faults. Another letter of the Bristol Wallsend Collieries Ltd of 1910 says:

'The original workings were not extensive and contained water and were from a previous company. There was an interval between the two company's workings during which the mine drowned out'.

The Bristol Wallsend Collieries Limited were clearly sold a concern which had never been viable and could have never made a profit due to adverse geology.

The Bristol Wallsend Colliery. This picture shows the last period of working about 1905, the simple headframe and single pulley was not unusual in small colliery at this time. The winding engine was a single cylinder engine with a large fly-wheel which was probably scrapped after the First World War, as your author was told that the local children played there at that time. These were some of the smallest trams to be seen; the output must have been small from the Mangotsfield seam.

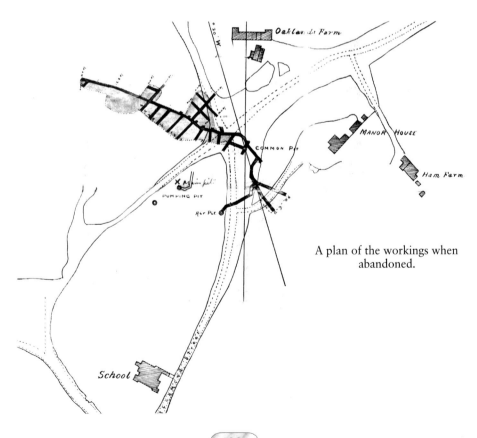

A plan of the workings when abandoned.

The Bristol Wallsend Collieries, Limited.

Colliery, MANGOTSFIELD, GLOUCESTERSHIRE.

TELEGRAPHIC ADDRESS
"TCHBLENDE, LONDON."

TELEPHONE
11845 CENTRAL

2, CHURCH COURT,
CLEMENTS LANE,
LONDON, E.C.

21st May 1909.

Sir,

I beg to give you formal notice that it is the intention of the above Company to abandon their Colliery at Mangotsfield which has been closed down since the 30th November 1907 and has been left in the charge of a watchman since that date.

All the pits have been securely railed off with barbed wire and everything else made safe.

Kindly let me know whether you wish to go over and inspect the fencings etc round the pits before I dismiss the watchman.

A copy of the plan showing the underground workings as surveyed just previous to the closing down has been made and will be sent, if you will let me know, either to you or direct to the Home Office.

Yours truly,

[signature]

Manager.

J.S. Martin Esq.,
  H.M. Inspector of Mines,
    16, Durdham Park,
      BRISTOL.

An interesting letter dated 1909. It is possible that the owners always hoped that someone would purchase the colliery one day, hence the reason for not filling the shafts and keeping the winding engine on site for some years after abandonment.

Church Farm Colliery

The original sinking date of this colliery is unknown. In the 1860s the concern was worked by Richard Haynes and Charles Emett. In 1870 they formed a new company which was known as the Mangotsfield Collieries Company Limited. They worked two pits, the Deep and Land Pits. In the 1970s the north and west showed signs of early workings in the form of crop workings and bell pits, and when the area was later developed for housing, areas of extraction were found, although many of the bell pits were extremely limited in extent and as they were filled they presented no threat to the properties.

In 1881 a Cornish engine was erected on the Deep Pit, where it worked 14in pumps with a 9ft stroke working at 7 strokes a minute, from a depth of 255 feet. The foundations of this were cleared in 1980, and found to be for a single cylinder winding engine. There were signs of rebuilding and clearly the original engine had been removed and a winder installed by the later company. The power was supplied by two egg-ended boilers.

In its final phase the two shafts were 420 feet apart. At Deep Pit the winding and pumping shaft was oval and 255 feet deep, with Land Pit being 78 feet deep to the Great Vein. The Little Vein was 15 feet below the Great Vein and the thickness of the seams were 2ft 8in and 1ft 10in respectively. The area of the colliery was 41 acres for which a rent of £30 per annum was paid and a further 16 acres for which a dead rent of £6 per annum was payable. A royalty of 7d and 6d per ton was paid on all coal raised. The lease was for 99 years. In 1870 it was claimed that the two seams could be worked at a rate of 50 tons per day at a cost of 5s 6d per ton delivered into carts at the pit bank. The roof was said to be strong, little timber was needed and the coal was free from faults. However the colliery closed in May 1891, with the lack of capital the reason for closure on the abandonment form.

A close inspection of the geology shows that this mine had little scope for expansion as large faults existed north and east, and to the south and west the seams were worked right up to the outcrop. There was also a small mine worked by Fryer and Co and known as Pomphrey Colliery 100 feet north-east of Pomphrey Farm. This had a steam engine and a horse gin and worked four seams from a shaft 42 fathoms (252 feet) deep. The mine had a short life and was certainly at work in 1841.

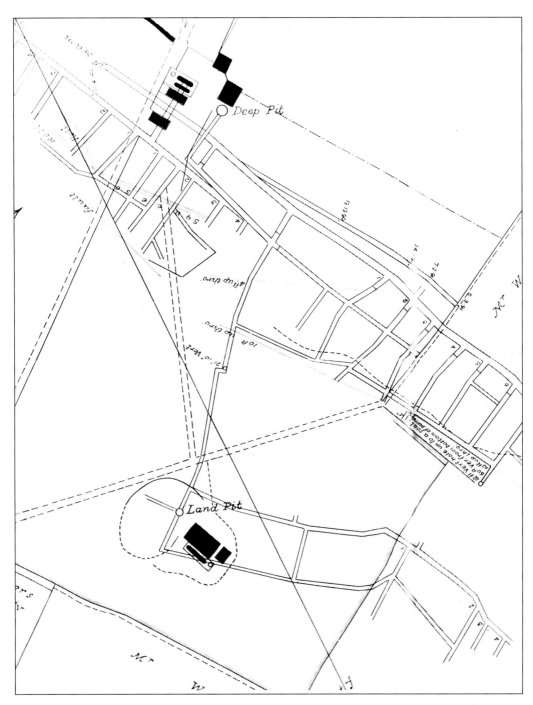

Plan of the workings of Church Farm Colliery after closure. Had the pumping been placed further to the deep a greater area of extraction could have been achieved.

The pumping engine house erected in 1881.

The colliery taken in 1980, the house in the background was Church Farm.

Parkfield Colliery

Parkfield Colliery was developed to the west of an old area which had been worked in the 18th century, from a number of abandoned pits which were mainly situated along the eastern outcrops of the Coalpit Heath basin.

The shafts (and where known, the depths) were:

Garden Pit	420 feet deep
Dudley Pits	345 feet deep
Cooks Pit	390 feet deep
Upper Wood Pit	?
Great Cart Pit	450 feet deep
Wood Pit	?
Old Engine Pit	450 feet deep
Bryants Pit	?
Shortwood Engine Pit	570 feet deep

The sinking of Parkfield Engine Pit commenced in May 1851, using sinkers from Staffordshire. It was not until 1 January 1853 that the first coal was reached, but it was a thin seam and was not worked. The coal at Parkfield was extremely good and highly bituminous and was mainly used for house coal and gas manufacture. The seams in the Parkfield take were very regular and totally free from faulting. The Parkfield shafts were sunk 831 feet to the bottom landing, plus a sump of 9 feet. Only the Upper Series were worked:

The Hard vein	2 ft,
The Top vein	2ft 4 in, and
The Hollybush	2ft 6in
The Great veins	3ft respectively.

The last two seams were separated by several feet.

In 1856 a collier, Job Rolph aged 21, said to have been under the influence of drink, went in search of a sledge hammer into workings which had been stopped for several weeks. The ventilation had been so arranged that whenever the water rose in the sump the current of air was cut off from the inner workings, and they were filled with carbonic acid gas, or chokedamp. The workings into which Rolph went had been standing full of this noxious gas for nearly three weeks. As soon as Rolph's absence had been noticed and his fate realised, Richard Wheeler, a married man with six children dashed up the heading with his jacket over his head but he was also overpowered,. The men could be heard groaning for some time, and the bodies were not recovered for some 18 hours. It is surprising that a heading in a new mine should be so affected. It must be that the heading was near or connected to Cooks Pit old workings, and it is known that Cooks Pit was reopened. This accident might have the result of this reopening.

The sale particulars provide a most detailed description of the collieries then with their machinery, buildings and mineral leases. At Parkfield Pit were a pair of horizontal direct-acting steam winding engines with 28in cylinders, 4ft stroke, and drum of 15ft diameter. This engine was recorded by George Watkins in 1932 and was noted to have been made by Teague & Chew of Cinderford in the Forest of Dean. There were two pairs of headgears 38ft high with 2 pulley wheels 15ft in diameter. There were also two wrought iron cages 11 cwt each. Steam was provided by four Lancashire boilers 27ft by 7ft. The Cornish pumping engine had a 54in cylinder, with a 7ft stroke, with a 12 inch pump, and wooden pump rods and cast pipes. The Cornish Pump had 2 Lancashire boilers. The ventilating fan was 18ft by 7ft, and driven by a pair of horizontal engines with 14in cylinders with a 16in stroke and compound slide valves.

One cylinder was a spare for use in case of failure of the other. There was also a surface beam engine with an 18in cylinder and 3ft stroke, and flywheel, 4 gearing wheels and barrel drums. This engine drove an endless haulage system down the shaft with a galvanised rope 990ft long and $1^{1}/_{2}$ in diameter, a second steel rope 990ft long, 1in diameter. There was a fitting shop, saw mills, forage stores, gas house and gas retorts with a gasometer. Underground there were 3 engines for haulage, 5,250ft of single T-headed rails, 4,350ft of bridge rails, 5,400ft of tram bridge rails and 4,800ft of sleepers and dogs with 208 iron trams.

Parkfield Colliery worked for another 36 years producing good quality coal from the Kingswood Great Vein but like many other pits with a number of fatalities, particularly under Handel Cossham. By 1936 water was becoming a problem and with increasing pumping costs and the decreasing reserves of coal the colliery became uneconomic. In 1936 the East Bristol Collieries Limited decided to close the pit from 15 August that year. The area of coal extraction at Parkfield was, by Bristol standards, large, the underground layout was well planned and little coal was left, unlike some other pits where large areas of coal remained unworked due to bad planning. The take was $1^{3}/_{4}$ mile long stopping just short of the Kidney Hill Fault in the north. To the west a major fault was the western boundary of the Parkfield workings, but no faults were known in the take. In the south the workings came almost to the outcrops. They were one mile wide commencing just short of the outcrops in the east and ran for one mile to the west at Lyde Green.

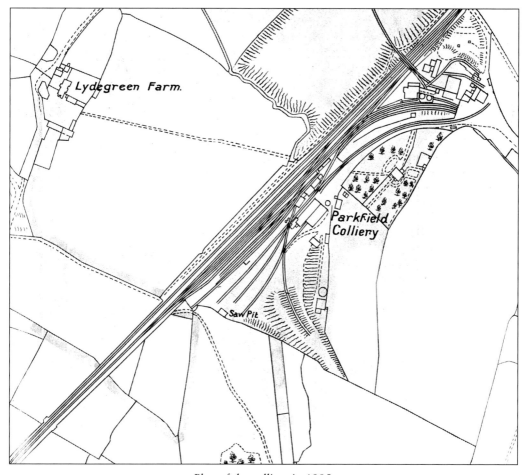

Plan of the colliery in 1895.

Brandy Bottom Colliery

As with many collieries the sinking date is not known, but the plans of 1840 give the impression that the pit had been working for some considerable time and must have been sunk in the 1780-90 period when the earlier workings up on the hill at Engine Bottom were exhausted. There is some confusion over the origins of the two shafts, but plans of 1840 show Lord Radnor's Pit with only one shaft, and a later plan of 1853 again only shows one shaft, this time known as Brandy Bottom Colliery, but later plans dating from the 1870 period show two shafts. On the 1853 plan, the age and extent of the workings are clearly shown, the inclines levels and general layout are poor and show little planning: main roadways suddenly turn and double back on themselves, needing many haulage engines or making the horse road very tortuous. Indeed one plan states that the Brandy Bottom South Pit is 480ft deep and the workings very old, probably 100 to 200 years. In 1853 an interesting feature is shown: an underground horse gin situated at the head of a long steep incline which is over 1,000 feet in length. It is highly likely that Cossham sank the second shaft, which was known as the New Pit and is shown to be 618 feet deep.

For many years the colliery was worked by Jefferies Walters & Co. who retained the pit from the 1850s until Handel Cossham acquired it in 1871. A close examination of the remains of the pit shows two phases; the heapstead can be seen to have been raised as the original work is in White Lias stone and some of the later phase is built in brick and Pennant sandstone. The Cornish pump looks old and must have been installed early as the Upper coal measures in this area are very heavily watered, and no pit would have worked in this area without substantial pumping capacity. The rail connections in 1883 show that some of the coal was being wound at Brandy Bottom in the 1880s. In 1900 the Cornish engine had a 60in cylinder which worked

Brandy Bottom Colliery in the early 1980s with the Cornish pumping house in the foreground.

with an 8ft stroke, the top lift plunger pump was 11in, the length of pump rods was 780ft and mainly 10 inches square. This engine had two Lancashire boilers 27ft long by 7ft diameter. The horizontal winding engines had 12in cylinders, 18in stroke, 8ft flywheel, two gearing wheels and barrel drum 7ft in diameter. The headgear for this engine had two pulley wheels each 7 feet in diameter. There were two cages made of cast-iron weighing 7cwt each, plus a little donkey engine with 7in cylinders and 5in stroke. The pit bank was 50ft by 26ft. At the New Pit there was also a condensing winding engine with 24in cylinders, 4ft 6in stroke, 2 gearing wheels, fly wheels and barrel drum 13ft in diameter. There was 1,140 feet of $1^{1}/_{2}$ inch steel wire rope, a wrought iron cage and a headgear with 13ft pulley wheels. The pit bank was shown to be 40ft by 40ft.

It may appear odd that the northern shaft at Brandy Bottom is known as the South Pit. This was for a time the southern shaft of the Parkfield Colliery complex after the New Pit was sunk sometime after 1871, and the original name was retained as it appeared on all the early mine plans. By 1899 The South Pit or Brandy Bottom Colliery was retained only for pumping and ventilation and the site remained as a pumping and ventilating pit. Later a single inlet Sirocco fan was installed in the New Pit. Some of the sidings were removed by 1903, and the rest were later lifted and the whole complex closed in 1936.

The South Pit heapstead in the late 1970s, mainly built from Lias with brick arches and quoins.

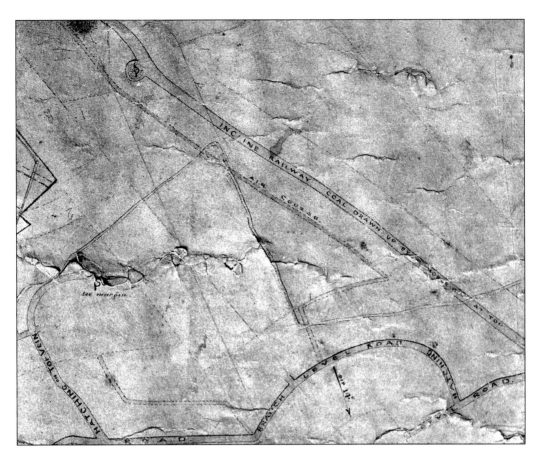

An underground plan of workings in the Top vein at Brandy Bottom Colliery dated 1853 showing a horse gin installed underground at the head of an incline to draw trams up the incline. Note the meandering air course, ventilation must have been very poor as the furnace was not large and had to ventilate a very large area of workings.

The Shortwood Collieries

Shortwood Collieries comprised a number of pits mainly dating from the 18th century, some of which were reopened by Handel Cossham. But as with many pits in this coalfield the origins of the Shortwood Collieries are obscure. In the case of the Upper and Lower Wood Pits a sinking date of pre-1800 is suggested since a section of the Wood and Upper Wood Pits, dated 1853, shows an extensive area of working suggesting working going back to the early 1800s.

The two engines shown on the section must be the earliest detailed drawings in the coalfield of a pit with their winding engine houses. These engines with their external winding drums are very early and must date from the 1820s. A similar drawing exists for the Durham coalfield and can be dated to around 1820. The oldest of the group of pits to the south of the old brick works are Chafffhouse Pit which is shown on a plan of Brandy Bottom dated 1840 as the sinking pit. No other pits are shown on this plan except Brandy Bottom and Wood Pits. Plans of Chaffhouse show that it was sunk to the Hard vein at a depth of 300ft and was finally abandoned in 1907. Cook's Pit situated south of the brick works should not be confused with the Cook's Pit at Parkfield; this late pit was also sunk to the Hard vein, Great vein and Top vein, with limited workings in the Hollybush vein, all abandoned by 1905.

Lapwater Pit was originally known as Shortwood Lower Pit, the depth on abandonment plans is 330ft. Limited workings are shown on the Top vein and Hard vein, but the early workings are only vaguely shown. All of the late Shortwood workings were abandoned by 1907, and were mostly worked by small single cylinder winding engines with large flywheels.

At the present time there is no evidence of earlier workings. The first reference to owners is in 1841 when the collieries were worked by Messrs Waters and Reynolds, who employed 120 hands, 30 of whom were boys, some only 8 to 9 years old. They worked the Top vein 3ft thick, the Hollybush vein 2ft 6in thick and the Great vein 2ft 8in and 20in thick. Wheeled tubs were used when possible. Coal was hauled from the gug incline by a windlass; 4 cwt of coal in a tub was drawn on a plateway.

By 1854 the Pucklechurch and Shortwood Collieries were now owned by Wethered, Cossham, & Bendall who in a set of rules clearly said that no boy under the age of 10 was allowed to work in the mine. The rules were very advanced for the time, with sensible suggestions like all stalls must be kept filled up to within 10 feet of the working face. Main airways must be contain no less that 8 square feet and any falls must be immediately cleared out. The Bailiff was to examine all air-courses once a week and to see that they were kept properly open for the ventilation of the mine; also to see that water-courses were kept properly open for the effectual draining of the mine. Any workman receiving injury was to be taken home with all possible despatch. Despite these rules, Cossham's pits had an appalling safety record. On 6 December 1852, 2 colliers, Eli Rogers 33 who had a wife and 2 children, and William Davies, single, were drowned in the shaft by water from old workings. The mine inspector commented that there were no plans and no precautions were taken; this again suggests old workings which were beyond recall.

Today apart from the sites of Lapwater and Chaffhouse (Shortwood Pit) nothing remains as clay pits and landfill have changed the landscape; even the impressive Shortwood Brick Works and manager's house are demolished and just Brandy Bottom Colliery still stands - soon to be restored.

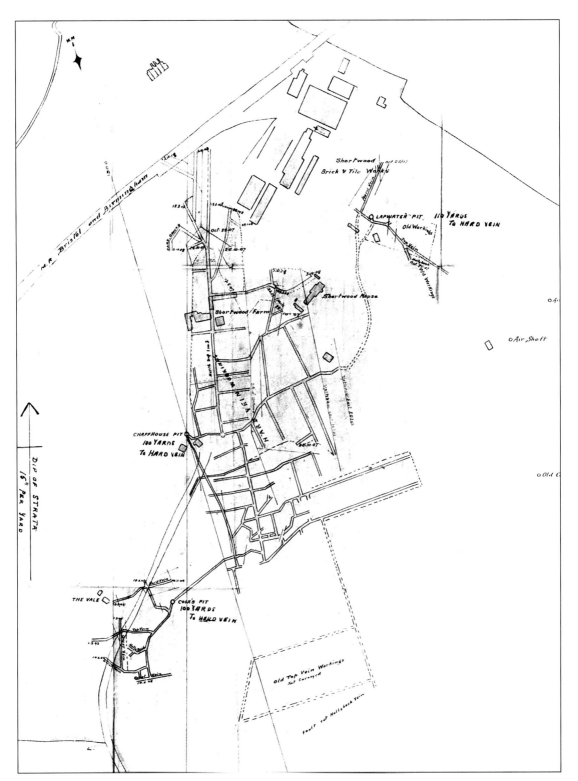

A plan of some of the workings of the Shortwood Collieries on abandonment in 1909.

The Wood Pit and the Upper Wood Pit in 1853 showing the old style engine house with external winding drums. Note the horns on the drum at Wood Pit which prevented the flat rope from slipping off the drum. (See over-page)

New Engine Pit and Orchard Pits

Ram Hill Colliery was sunk around 1820 to 1830 and was 558ft deep. It was originally worked by horse gin, but later by a beam winding engine, the remains of which were visible. This colliery also closed in the late 1860s the same time as New Engine, Churchleaze and Orchard Pits. The engine house of New Engine Pit still stands as did some of the workshops in the 1980s. After abandonment of the pit, it became workshops and saw mill for the Coalpit Heath Company. The remains of an egg-ended boiler which drove the old beam engine from the pit are situated in the old locomotive shed near Bitterwell Pond and were used as a water tank for *Lord Salisbury,* the Peckett 0-6-0 saddle tank which was eventually taken to Norton Hill, at Midsomer Norton after Frog Lane colliery closed. Today nothing remains of Churchleaze Pit; a few fragmentary walls were still standing at Orchard Pit, but the best preserved site is that of Ram Hill Colliery. A curious pit is the Dudley Pits of which very little is known. It was originally sunk in the late 18th century, but latterly owned by the Coalpit Heath Company, who probably sank the second shaft which had a horse gin and steam engine. This pit also worked until the 1860s and closed with other smaller Coalpit Heath Company pits as Frog Lane Colliery began to increase its extraction.

The remains of the New Engine egg-ended boiler in the engine shed near Bitterwell Lake. The end of the boiler was used as a water tank for the steam locomotives of Frog Lane Colliery. A fire kept the water from freezing in cold weather.

The Collieries of the Golden Valley

The coal seams of the Bitton District belong to the Upper Series. South of Redfield Hill is a small area of exposed coal measures about 1 mile in length. In this area four of the seams of the Upper Series outcrop. Some of these seams outcrop in the field behind the Golden Valley Old Pit in the following order north to south.

    Parrot vein
    Buff vein
    Rag vein
    Millgrit vein

These seams which also worked from Hole Lane Colliery.

The Golden Valley seams dip south-west, with a dip of 12 inches to the yard or 1 in 3. 1200ft to the south of the Old Pit is the Bitton Fault which runs in an east-west direction. This fault, which has a down-throw to the south, is also the southern limit of the workings from the New and Old pits.

Shaft Section of Golden Valley Old Pit

Seam	Thickness	Depth
Millgrit Seam	2ft 6in	26ft
Rag Seam	2ft 0in	74ft
Devil's Seam	not known	282ft
Buff Seam	not known	300ft
Parrot Seam	1ft 9in	421ft
Little or Brimstone Seam	not known	456ft
Muxen Seam Bad	not known	not known
Coking Coal	2ft 0in	774ft
New Smiths Coal	2ft 0in	938ft
Kenn-Moor Seam	2ft 0in	980ft

The shafts of both the Old and New Pits were sunk through 600ft of Pennant rock, which included the Millgrit, Rag, Buff and the rock covering the Parrot seam.

Shaft Section of the Golden Valley New Pit

Seam	Thickness	Depth
Coal measures (Pennant grit, with thin Fig and Francombe seams)		504ft
Millgrit Seam	2ft 6in	506ft
Rag Seam	2ft 0in	554ft
Devils Seam	unknown	762ft
Buff Seam	unknown	780ft
Parrot Seam	1ft 9in	901ft
Little or Brimstone Seam	unknown	936ft

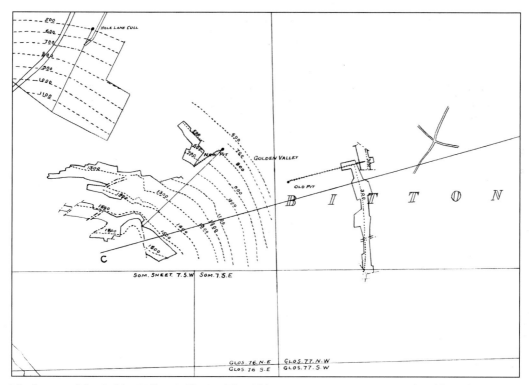

The layout of the Golden Valley Collieries. The Old Pit as its name suggests was the oldest, the New Pit being sunk in the 1820s. The main workings of the New Pit were in the Parrot vein which were only eighteen inches thick and 1,920ft deep.

On the Old and New Pits, the seams were followed to the dip for a distance of about 3,000 feet until the Bitton fault was reached; this fault was the limit of workings in the Golden Valley.

The colliery was worked to a considerable depth, 2000 feet, in fact, very few collieries in England worked such thin seams, at such depths at a profit. The depth of the New Pit shaft was 936ft, the deep workings are reached first by a 2,400ft incline from the pit bottom, worked from the pit bank by the winding engine. Then came an incline 150ft in length worked by a wheel and rope, followed by an incline similar to the last 150ft in length, then another incline 180ft length, up which the puts were dragged by manual labour. There appears to have been a similar arrangement at the Old Pit which was probably in use as late as the 1870s, when the reworking of the pillars was taking place in the Old Pit.

The earliest evidence of coal mining in the Golden Valley to date is a sale document dated 1726, for various plots of land, one of which is the field later known as the Boyd Field in which the ventilation furnace now stands.

A section of the document states:
"Excepting and always reserving out of these presents and the Grant hereby made unto the said Thomas Edwards his Heir Assigns all Mynes and Veins of coal and full and free liberty for him the said Thomas Edwards his Heirs & Asssigns from time to time and at any times hereafter to enter into and upon the said Premises or any part there of and digg and land Coles and such Coles with Carts Carriages and Horse to carry away and sell and dispose of at pleasure to his and their own use.

And likewise to make Dreins and Aqueducts for carrying the water in and through the said premises or any part thereof the said Thomas Edwards paying reasonable damages for spoil of the Grass or Corn there growing".

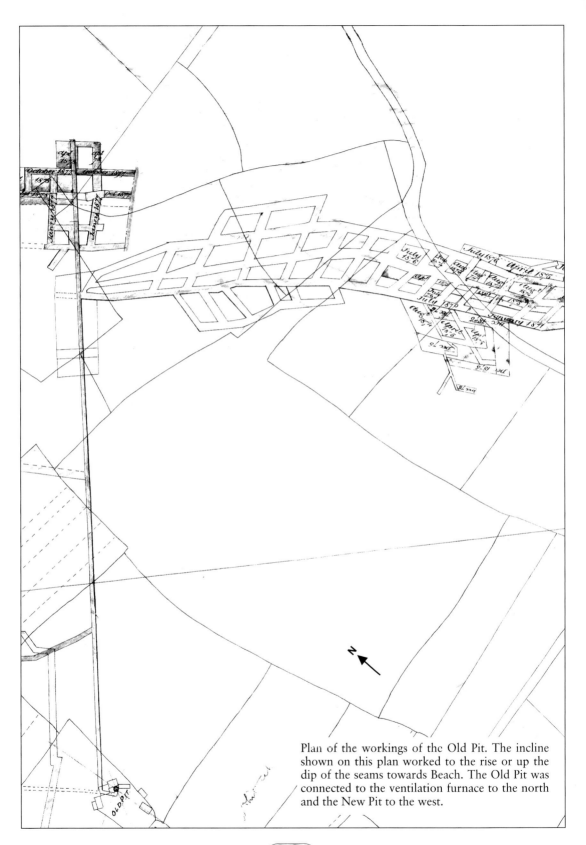

Plan of the workings of the Old Pit. The incline shown on this plan worked to the rise or up the dip of the seams towards Beach. The Old Pit was connected to the ventilation furnace to the north and the New Pit to the west.

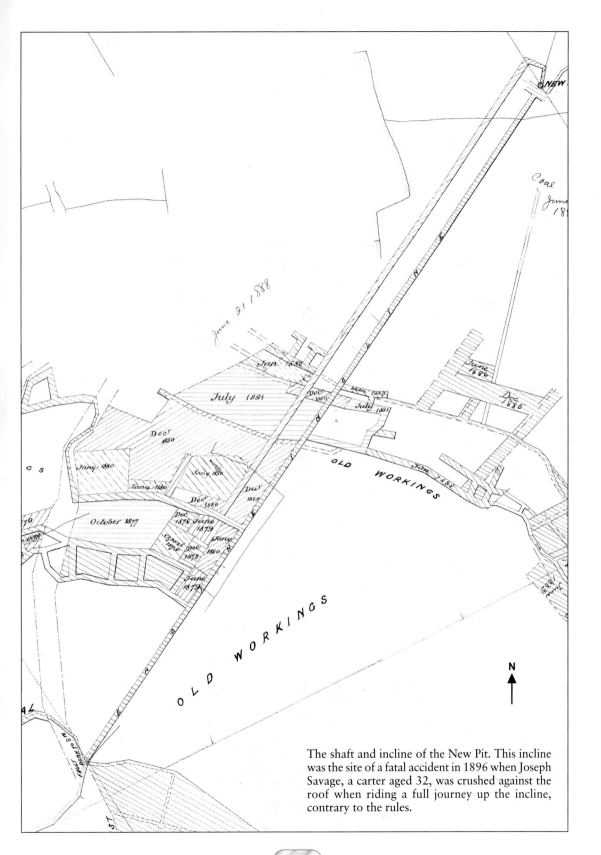

The shaft and incline of the New Pit. This incline was the site of a fatal accident in 1896 when Joseph Savage, a carter aged 32, was crushed against the roof when riding a full journey up the incline, contrary to the rules.

From this document we can assume that mining has already begun and that pits were working, or were about to work with water power, hence the 'aqueducts and dreins for carrying water in and through the said premises'.

There is a document in the Gloucester Record Office dating from 1814 which is part of the Sherwood Manuscripts, referring to the remains of an old coal pit which was converted into a boring mill in 1775. This old pit had a water wheel which pumped water from the mine.

Robert Newman stated that he:
'hath known the place where the Plaintiffs Mill is erected upwards of 70 years, that it was then a Coal Pit -that the wheel was against the hedge, that about 44 years since one Barnes used to land some lime coal there - that Barnes filled up the holes - that Witness rented grounds near there. That there was a little house which was used as a Counter to keep an account of the coal'.

It is now clear that early pits of the Golden Valley were situated on the exposed section of the coalfield very near to the River Boyd, so as to be in a position to use it as a source of power for their water wheels. By the l750s, the pits had moved away from the lower part of the valley and the River Boyd and were working on and around the outcrops high on the hillside using the cog and rung gins. One of these sites was uncovered by an excavation for a gas main which unfortunately destroyed the site. To date 10 shafts have been located, including the Old Pit shaft in the area to the south of Redfield Bridge. All were sunk between 1700 and 1800.

In the First World War when extra ground was needed for cultivation, the spoil heaps from one or two of these early sites were removed and deposited on the site of the Old Pit. It soon became apparent in the early stages of the archaeological excavation of the Old Pit that spoil from the other older sites had been dumped and spread about there, completely covering what remained of the site; the engines and boilers had been broken up for scrap before the First World War. It has been possible to date these pre-Old Pit sites to 1750 from clay tobacco pipes found in the spoil.

On the 3rd September 1798 a partnership was drawn up between Aaron Brain and James Quarman, Edward Stone, the elder Edward Stone, the younger Edward Stone, Samuel Brain, John Brain, William Lacey and Stephen Matthews. All except Stephen Matthews were signatories of a lease dated 19 July 1798, the lessors being George Flower, of Saltford, gentleman and Mary Flower of Bitton, widow.

The lease is in the Gloucester Record Office and consist of two pages 30in by 26in, the interesting part stating:

'... and grant unto the said Aaron Brain and William Lacey Full And Free leave Liberty And Licence for them the said Aaron Brain and William Lacey their executors administrators and their agents workmen and servants to enter into and make and open one or more new pit or pits and out of each pit or pits to raise and land any coal that is lying or abiding under ground in six certain pieces or parcels of pasture land of them the said George Flower and Mary Flower called or known by the several names of The Sheppards Leazes, The Three Acres, The Long Leaze, The Five Acres and the Land-furlands situate on the Upton side of the Brook or river at or near the Mill now in the possession of the said Mary Flower...... and to draw any level or levels and to erect a fire engine or engines for draining and carrying away all such water as shall or may annoy obstruct hinder the working of any mines veins or seams of coal as shall be so found in upon or under the ground aforesaid...'

In 1808 the first entry in the only surviving minute book refers to a meeting of the Colliery Proprietors on the l8 April l808, who asked Aaron Brain to: 'Look out and Purchase from 20 to 40 fathoms of second hand slides and rods which are expected to be wanted for the prosecution of the work'.

This is clear evidence that shaft sinking was still in progress in 1808, so it is unlikely that the pumping engine had been installed by 1808. It is highly possible that water was raised in a kibble or hudge by a horse gin. Also at the same meeting Aaron Brain was also, 'empowered to purchase in the best manner possible a second hand boiler of such dimensions as he may deem proper'. Perhaps the second hand boiler was for the pumping engine or used as a spare or standby boiler.

At a meeting on the 11 July 1808 it was: 'Resolved- on the recommendation of Aaron Brain that the Whimsey should be removed at a convenient time from the present situation to the upper pit in Mr Ferris's ground'. Mr Ferris's ground is not known, perhaps it was one of the pits situated on the hillside between Upton Cheyney and the Old Pit. At this point the entries in the Minute Book cease for 11 years.

On the 3rd May 1819 the proprietors resolved: 'that the Company do purchase the Atmospheric Engine now standing at Staple Hill.' This engine belonged to Mr Peterson and Mr Boult. The company paid £113 exclusive of House, Spring Beams & Shears. The minutes also tell us that: 'in contemplation a new company is about to be formed at Staple Hill. It is agreed to offer that Company when formed the use of the engine, if it is desirable to them at £20 per Ann. on their engaging to give it up on having 6 months notice from the Golden Valley Company'.

Although the proprietors do not say where the atmospheric engine is to be erected it was very probably intended for the Old Pit. It is unlikely to be the New Pit where sinking commenced about 1823 and took 7 years to complete. At a meeting in October 1821 Aaron Brain told the proprietors 'that a Whimsey of greater power is necessary for the work and will greatly benefit it'. It is very likely that the shaft of the Old Pit was deepened at this point in time.

As the New Pit did not reach depths where a steam winder would be needed until the mid- or late 1820s, this whimsey must have been intended for the Old Pit, to replace an engine with less capacity.

It appears that the New Pit came into production in 1830 or soon after and became one of the two downcast shafts for the Golden Valley complex. The Old Pit then became a pumping pit, and a second exit for the New Pit, although 30 tons of coal per week was still raised at the Old Pit as late as 1870, possibly for the boilers and ventilation furnace. On the 23 December 1834 between 5 and 6 in the morning, a number of colliers descended the shaft of the New Pit, which was 936ft deep. After descending several feet the rope suddenly snapped, precipitating the men 900ft down the shaft. Four men were dashed to pieces, while the other four were so lacerated it was felt that there was little hope of their survival.

Miraculously a boy named Daniel Harding, and a collier, Joseph Bawn, grabbed a chain which hung in the shaft to serve as a guide. Once the men on the pit bank realised that men were clinging to the side of the shaft, a man was sent down with a rope with a noose to render assistance. The man descending the shaft then came across Daniel Harding first who cried out 'Don't mind me, I can still hold on a little longer, Joseph Bawn who is lower down is nearly exhausted, save him first.' The man was lowered further and found Joseph Bawn as described by the boy, and after bringing him up safely, again descended, and succeeded in bringing Daniel Harding safely to the surface.

It is recorded, that from the time of the accident until the boy was brought to bank fifteen to twenty minutes passed.

In May 1841 Mr E Waring visited the colliery on behalf of the Children's Employment Commission who were investigating the conditions and treatment of child labour. The underground manager, William Bryant, aged 41, stated that not more than eight boys under 13 were employed, the youngest being about 11. He also had six lads under 18 earning from 1s 6d to 12s a-week. All were carters, they required no special door-keepers, they worked a seam 2ft thick but it could be irregualr; they worked from five in the morning until one in the afternoon.

The report also stated that the foul air was kept under by ventilation, but was often troublesome, but unfortunately it does not tell us what system of ventilation was used, or where the furnace was. The under-manager said that one man fell down insensible from foul air, but was restored on being taken into fresh air. Mr Waring also examined William Short aged 11. He carted a bushel and a half up a ladder by the girdle; he said it did not hurt him then, but used to at first. Short's wages were 4d a day.

An accident occurred on Sunday 26 March 1882, when smoke was found to be issuing from the top of the New pit shaft. The bailiff and his son with another man gave instructions to raise steam on the winding engine at the Old Pit, which at that time was used for pumping and was also the second means of egress from the workings of the New Pit.

The bailiff intended to descend the Old Pit and travel underground to the pit bottom of the New Pit. When he reached the bottom of the shaft, the empty hudge was sent up to the surface, which was the usual custom as the shaft was wet and hemp rope was used. After half an hour the signal was given to lower the hudge, which was done immediately, but as no further signal was given to raise it again the hudge was left at the pit bottom. Sometime later the men at the top felt that something might be wrong. They had the hudge brought back to the surface, and then descended and found the dead bodies of two of the men close to the shaft, and the third body further into the heading. The men who first went down experienced a strange smell not unlike sulphur, their lights were put out, and they became giddy and had to return to the surface. Some water was turned down the shaft, and very soon after other men were able to descend and recover the bodies. The owner then closed down the shafts, with a view to extinguishing the fire. No one was then allowed to enter until the fire was thought to be out.

When it became possible to descend the pit, it was found that a large heap of coal and some timber near the bottom of the shaft had been on fire. It was thought that one of the workmen coming out of the pit on Saturday had thrown down the end of his candle, which had ignited some dry timber at the side of the road.

During the inquiry it was disclosed that the winding engine man was stone deaf. Mr. Brain advised that someone else should be appointed in his place. The jury also adopted Mr Brain's recommendation that Abraham Cook, senior, Abraham Cook, junior, and Alfred Walter, were accidentally suffocated in the Old Pit, Golden Valley. The last accident at the Golden Valley Colliery was in March 1897. It was fortunately not fatal. A workman had his foot crushed at the pit bottom by the cage alighting on it in consequence of his standing too near the shaft.

An old pit in the Boyd Field was reopened and repaired, and a ventilation furnace was built alongside the shaft sometime between 1840 and 1870. This shaft then became the upcast shaft for the Golden Valley Colliery. Along with its horse gin circle it is shown on the map of 1840, but the furnace does not appear to have been built until after that date. A lease dated 1877 describes the furnace as a Chimney and Cowl with an underground connection to the Golden Valley Colliery, but furnace and underground connection are not shown on any of the colliery plans of that period. The rent for the shaft and furnace in 1876 was £22 per year, this was raised a year later to £28.

By the late 1870s most if not all available reserves of coal were exhausted, and the production for the last 20 years of the pit's life was augmented by robbing the pillars. The output of coal for the last year was 8,969 tons which is approximately 175 tons per week. The workforce was thought to be 70 or 80 men and boys.

The colliery was offered up for sale in February 1898, a notice on the front page of the *Colliery Guardian* stated:

'This Old-Established Colliery that has been worked more or less for upwards of 60 years, and is well known in the market as an Excellent Smith's & House Coal.

To a Capitalist prepared to spend a moderate sum in developing The Parrot Seam eastwards and the House Coal to the deep, a very profitable Colliery should be easily established.'

According to the late Tony Brain, the colliery did not reach the reserve price and was withdrawn, and then sold off piecemeal. The mining rights were purchased by Phillip Fussell of Hole Lane Colliery to prevent any encroachment by the Golden Valley Colliery. It is very obvious that the Golden Valley Colliery was completely worked out by the 1870s, but how the colliery managed to keep working until 1898 is a mystery. The plans show the company had robbed most of the pillars in the Parrot vein and New Smiths Coal seam. Development headings had been driven in an unsuccessful search for unworked coal, but most headings ended in a vertical borehole, which in most cases found little or no coal. Unknown to the colliery proprietors, on the south side of the Bitton Fault lay many thick seams of high quality coal, including the Doxall, the Toad vein, Kingswood Great vein and the Ashton Series of seams. In fact a new virgin coalfield lay to the south of the Fault.

In the early 1980s the site was excavated by a group of industrial archaeologists who uncovered the remains of a small engine house for an atmospheric pumping engine, which was 14ft wide and 18ft long, but most of the masonry had been robbed out. The bob wall which had been built from the best masonry stone, was robbed down to a height of 8ft on the eastern side and to mostly ground level on the western side. The shaft which had been filled with material from the buildings on the site, was rectangular with rounded corners, typical of the 1790s, and was 8ft by 5ft 6in. The shaft would have been divided into two sections by a brattice, with pump rods and rising mains in the northern section and the winding by a hudge in the south section.

Foundations of a large well-preserved haystack boiler and a small egg-ended boiler were uncovered on the western side of the stump of the pumping engine house. The colliery winding engine was a typical rotative beam winder of the early 19th century, which was enclosed in a structure slightly smaller than the pumping engine house. The pit for the winding drum or reel, was 16ft by 9ft, which does suggest that in the last phase of the engine's life a drum was probably installed. Very little remained of the winding engine house: no walls had survived, only the hot well and foundations, and holding-down bolts for the cylinder remained, with a large drum pit. A rare and interesting find was the remains of the underground haulage system which was driven by the colliery beam winding engine. The remains of the system consisted of a slot which carried the endless rope or chain over a large pulley situated over the edge of the shaft, and down the shaft, to move trams up an incline from the pit bottom into a district where coal was worked.

The engine appears to have had a 14ft long beam, a cylinder of approximately 30in diameter, with a stroke of about 4ft. The diameter of the flywheel was 12 to 13ft. This engine was driven by a wagon boiler, but it was impossible to say whether the engine was of the atmospheric type or not, since these later atmospheric engines had pickle pot condensers, and the foundations of a single acting Boulton & Watt type engine would have been similar. In the excavations a brass tap was found, and although it cannot be certain that this is the tap controlling the water supply to the open top cylinder, there is a chance that it could be, as there was an extremely worn brass bearing also found in the engine house. Alongside the top of the shaft, one cast iron link forming part of the chain which connected the beam to the wooden pump rods was found.

An interesting structure uncovered was the rope run from the manually operated capstan, which was used for changing or repairing the pump rods in the shaft. This channel which is 2ft wide and 2ft deep ran from the sheer-legs positioned over the southern section of the shaft in a westerly direction. The sheer legs needed enough height to allow a full length of pump rod or rising main to be drawn vertically from the shaft. Unfortunately the site of the capstan is thought to be under a large oak tree on the edge of the site, and could not be investigated.

The excavations of the Old Pit and the restoration of the ventilation furnace were unique, they were probably the only large-scale excavations at the time to uncover and record a pit of the 1790-1810 period. Unfortunately the Old Pit is today very overgrown, as the site and the ventilation furnace are both on private land with no public access.

Early Collieries at Coalpit Heath and Westerleigh

The working of coal was well under way by the 1680s although far earlier workings existed before then. The earliest coal owner at Coalpit Heath was Samuel Astry, the Lord of the Manor of Westerleigh. His estate papers still exist in the Bristol City Archives. Samuel Astry died in 1704 and his wife in 1708. The estate then passed into the hands of Sir John Smyth of Ashton Court, when he married one of Astry's daughters. Later Smyth established a partnership with Lord Middleton and Edward Colston. The Middleton family still held shares in the Coalpit Heath Company as late as 1853, but by the 1870s only Sir Grenville Smyth is listed as a colliery owner. The Smyth family retained an interest in Frog Lane Colliery until 1947 and the nationalisation of the Coal Industry.

By the mid-18th century all three outcrops had been worked and in some places it is recorded that outcrops were reworked the second time before 1740. As in all outcrop areas many bell pits had also been sunk and drainage levels driven. The first steam pumps had been installed some time before 1769, as Donn's map of that year shows two engines close together. A slightly later map dated 1772 shows the Ram Hill Engine, with a second engine due west, this may be the Old Engine Pit and is likely to have been the first engine erected and as expected it is close to the outcrops. The Ram Hill and Serridge Engines went much deeper. The site of the Old Engine Pit is now under the railway line. The Ram Hill engine is highly likely to be the second engine and is sited 1,300ft to the deep and east of the Old Engine. The third engine at Coalpit Heath was at Serridge, and it was started in 1790. No records of the middle period give any locations but it is suspected that the Ram Hill Engine Pit was the first large pit with its coaling shaft 230ft to the west of the pumping shaft. Today the site of the pumping shaft is filled, but some 10 to 15 years ago the large well-made reservoir was uncovered. The coaling shaft was just a shallow depression in the ground. At Serridge Engine the tunnel giving access to the boiler alongside the engine is still complete although partially filled with rubbish.

From this time pits were sunk across the Smyth estates. Accounts of 1791, list No.1 and No.2 Churchlease, Serridge Pit, The New Pit (is this New Engine Pit?), Horses Engine Pit, and four other pits whose positions are now unknown, one of which may be Nibley Pit. But on a map which is undated, but thought to be around 1820, 20 coal pits are shown in Coalpit Heath, but it does not differentiate between working and abandoned pits. Many of these pits worked right into the 19th century; some like New Engine Pit remaining as a workshop until the closure of Frog Lane Colliery in 1949. Others like No.11 or Orchard Pit were situated to the rise of New Engine Pit and were comparatively shallow and worked mainly with a horse gin although reservoirs also suggest a steam engine. The drainage ran along a water level to Ram Hill pumping pit. Church Lease Pit situated to the north of Serridge Pit which only had a horse gin was certainly at work in 1820 and the removal of the pillars took place in the 1865 to 1870 period. This appears to be the last date of working.

The restoration of the base of the haystack boiler, Golden Valley Old Pit.

The base of the cylinder of the winding engine in foreground, Golden Valley Old Pit. The narrow slot at the top of the picture is for the flywheel with the pit for a winding drum and the endless haulage, all driven off the geared flywheel.

The slot for the chain or wire rope for the endless haulage running into the shaft which is capped with brickwork.

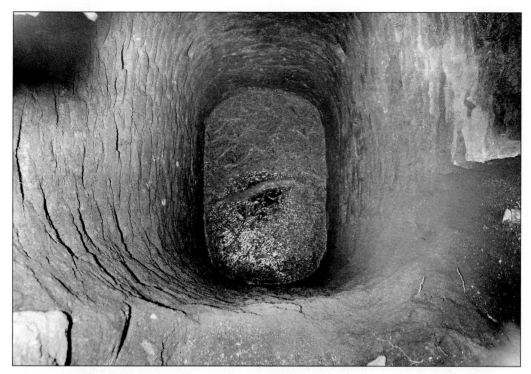

The shaft of the Golden Valley Old Pit. The shaft was filled by pushing the buildings down the shaft. The slot for the balance box mechanism of the pumping engine is on the right of the picture.

Base of the spare boiler for pumping engine. This had been an egg-ended boiler, the pipe providing extra draught for the firebox by supplying extra air under the fire bars. Very thick deposits of lime were found around the boiler area suggesting that the boilers were furred up and the extra draught was needed to raise sufficient steam.

Left: The excavation of the stump of the atmospheric engine house. Two holes in the back wall were the sockets of the two spring beams. Apart from a brass tap, nothing was found in the engine house.

Below: The Ventilation Furnace before restoration.

Above: The furnace during restoration. The platform for a horse gin was uncovered alongside the furnace. The gin was needed to inspect and maintain the shaft which was in use as an upcast until the colliery closed.

The furnace surrounded by scaffolding.

The restored furnace.

The Golden Valley, Bitton with the overgrown furnace before removal of the ivy.

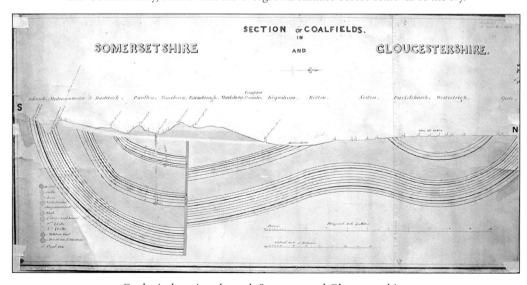

Geological section through Somerset and Gloucestershire.

Frog Lane Colliery at Coalpit Heath

This was the last of the old pits in the northern portion of the Bristol Coalfield and with exception of the Harry Stoke Drift Mine was the last, certainly the end of an era, bringing to a close hundreds of years of mining in the Bristol Coalfield. The earliest record for Frog Lane Colliery is the Inspector's report on it for the 19 July 1853 when George Flook, a sinker, overbalanced while he was plumbing the shaft. This colliery was sunk by the Coalpit Heath Collieries early in the 1850s, to the north of the workings of the early pits which were all situated west of the Coalpit Heath fault. There were three separate sites: Frog Lane for winding and pumping, Mays Hill pit for ventilation and Nibley Colliery for additional pumping. Nibley was an extremely old pit with origins back in the late 18th century. Frog Lane shafts were 660ft in depth for drawing coal and pumping water. The Mays Hill shaft to the High vein was 390ft in depth, it was 1,350ft from Frog Lane and used as a fan shaft. Nibley shaft to the Hard seam was 138ft deep, 750ft from Mays Hill Pit and was used as a pumping shaft. At Frog Lane Colliery there were two shafts, downcast, each oval in section and just 7ft apart. The winding shaft was 9ft by 6 1/2ft, and 660ft deep. The pumping shaft was 9ft by 6ft and 690 feet in depth.

The Hard seam was thicker at Coalpit Heath than at Parkfield Colliery. The High seam was known as the Hollybush and the Great Seam is united. The original winding engine at Frog Lane was of the vertical type similar to that of the Glyn Pits at Pontypool, which is now preserved. Later this engine was replaced by a horizontal winder with 2 cylinders 27in in diameter with a 4ft stroke, and a plain 12ft drum. This engine could draw 400 tons of coal in ten hours, winding two tubs on two decks in each cage. The cages ran on wooden guides. A compound hauling engine was placed in the High Seam, about 60ft south of the shaft. The engine had 15in and 26in cylinders and condenser; the stroke was 18in and the engine was geared 1 to 4 and had two 6ft drums. The engine hauled over a distance of 6,000ft with variable gradients. From this point the heading continued for a further 2,310ft, but then horses were employed in the level roads.

The engine, supplied by steam raised on the surface, could draw a set of tubs from the terminus in 7 minutes. The gauge underground was 2ft, with the main roads being 9ft wide by 6ft high. The roads were normally ripped two or three times until a good stratum was reached which formed a good roof. The pumping engine at Frog Lane was made by J. and E. Bush of Bristol in 1852; it had a cylinder $85^{1}/_{2}$in in diameter and 10ft stroke; it was later fitted with Davey's differential gear. It made six and a-half strokes per minute, and worked 10 hours out of 24 and the water was raised in three lifts. There was also a second pumping engine at the Nibley shaft which had a horizontal cylinder, 16in by 3ft stroke, geared 1 to 3 which worked two bell cranks fixed over the shaft, by means of a connecting road from the disk. There were two 8in lifting sets with a stroke of 3ft 4in, working 16 strokes a minute, working 10 hours out of 24. The Bush Cornish engine raised 2,246 tons of water per day, with the Nibley engine raising 616 tons per day, making a total of 2,862 tons of water raised per day.

Three coal seams were raised at Coalpit Heath: the Top or Hard seam, 2ft 6in thick, was a good strong coal which travelled well, and when burnt would leave a red ash: the Hollybush vein was not worked until after the last war as it was difficult to work and had a bed of fireclay and poor roof. The High vein was in two bands and the total section was 4ft thick, but not considered as good as the Hard vein. To work the Hard vein a horizontal cross-measures drift was driven from the High seam at the bottom of the shaft to the Hard vein in a straight line, cutting the latter at 1,500ft. Coal was got on the pillar and stall system. The wagon roads were 9ft wide by 6ft high.

The system in the 1890s was somewhat archaic. The stalls were driven parallel to the wagon road 10 yards wide and with 6 yards width of pillar between them; the cross hatching were 10 ft in width and the length of the pillar was 120ft. Usually three stalls were driven abreast of the wagon road for three pillars forward, when the coal left between the stalls was worked back. The coal was brought down the hatchings by tugger boys. The apparatus used was a chain fastened to a prop at the top of the hatching, a ladder was laid on the floor, and a wooden tub, carrying 3 cwt, which was contrived so as to run up and down on the ladder. The tugger boy pulled the tub up by tugging at the chain and lowered it down in a peculiar way, using the chain as a regulator where the seam lay at an angle of 20 degrees. The permanent roads were ripped two or three times for height, and powder was used. In temporary roads 2ft of top may be taken down in the High vein and 3 to 4 feet in the Hard vein, the coal being brought down by wedging. A Guibal fan was erected at Mays Hill Pit in 1873, which had a diameter of 16ft and was driven by a steam engine with a 12in cylinder and an 18in stroke. A duplicate engine was on standby, the engine and fan being kept working at a speed of 95 revs per minute. There was also a Capell fan at Mays Hill which had been installed in 1890. This fan was 12ft in diameter, steam driven, with the fan working at 128 revs per minute, with 38,000 cubic feet of air per minute being produced.

At nationalisation in 1947 the colliery employed 258 men underground, who produced around 3,500 tons of coal per month mainly from the Hard and High veins. As the colliery was nearing exhaustion attempts were made to try and work the Hollybush vein in 1944. Two headings were commenced in disturbed ground; the development was pursued in the hope that conditions would improve, but the two panels driven encountered rolls in the coal, and with the poor roof, the workings in the Hollybush were abandoned in 1947. Once the other two seams were exhausted the colliery closed in 1949.

As mentioned elsewhere, New Engine Pit yard was used by Frog Lane Colliery as the workshop, foundry, smiths and fitting shop with the saw mill driven by the old beam winding from New Engine Pit. One end of the egg-ended boiler is in the remains of the engine shed near Bitterwell Lake. This section of boiler was used to fill the boilers of the locomotives. The grate under the boiler was to prevent the water from freezing in the cold winters which were common at that time. Today the only structure of Frog Lane Colliery still standing is the steam winding engine house which has almost lost its roof.

Opposite page: Plan of the early pre-Frog Lane Workings at Coalpit Heath in the High vein. These workings were all situated to the west of the Coalpit Heath Fault, the later deeper workings of Frog Lane Colliery being on the eastern side of the fault. A pit adjacent to Ram Hill Engine Pit was probably the earliest sinking in this group of pits to be followed by Surrage now known as Serrage Pit, New Engine No.11 or Orchard Pit and Churchleaze Pit. Ram Hill Pit was the last sinking in this area.

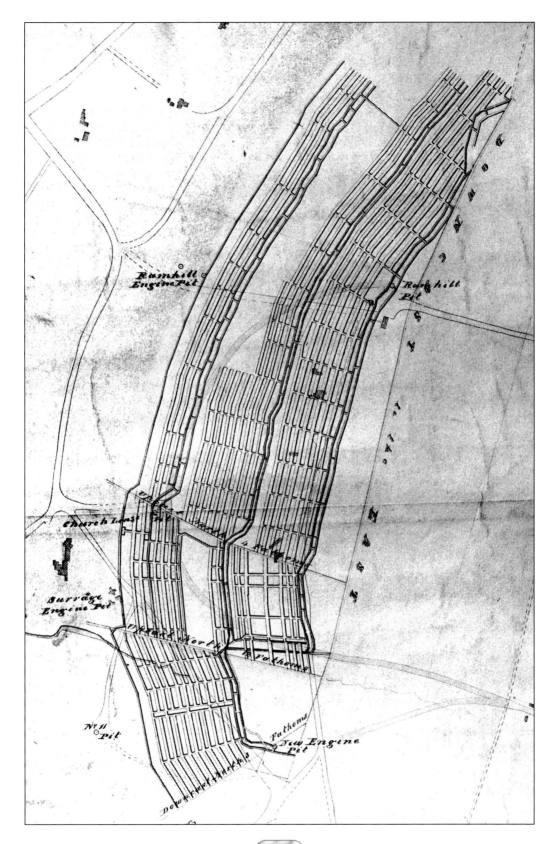

This is the plan of Ram Hill Pit at abandonment, showing the horse gin and shaft but not the steam winding engine which is known to have been in place at closure. The pillar of unworked coal around the shaft shows very clearly. The layout of this pit was well thought out, with the main roads running east to west along the strike of the seams, so that the horse roads are level. Horses were used by the Coalpit Heath Company.

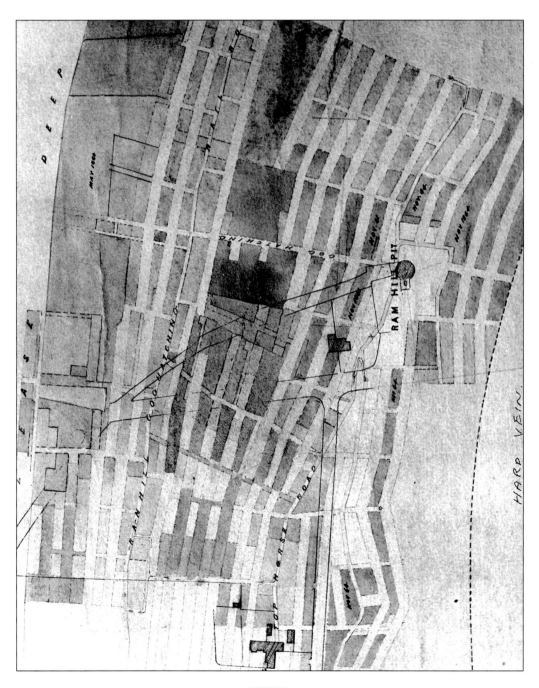

The oval shaft of Ram Hill Pit.

The excavated remains of the beam engine site at Ram Hill. This project was carried out by Avon County as a job creation scheme.

Opposite page: The terminus of the Tramway at Ram Hill. It was horse drawn with wrought-iron fish-bellied rail on stone sleeper blocks. The horse gin was situated behind the archway.

Serrage Engine Pit, remains of an arch and tunnel which could either be the archway into the ash pit under the haystack boiler, or the rope-way between the capstan and sheer legs of the pumping engine. The capstan raised the wooden pump rods or the massive cast iron rising mains if there was a problem.

Frog Lane Colliery taken soon after sinking, possibly around 1860-70. The tall engine house held a vertical winding engine with flat ropes. The Cornish beam engine remained on site until closure in 1949.

Frog Lane Colliery in the 1900-05 period. The vertical winder has been replaced with a horizontal winder.

A view of Frog Lane Colliery 1900-05 with the then manager Francis Eames in the white coat in the timber yard.

The pit bottom at Frog Lane c.1905, with hitchers at work with peg and ball oil lamps. The wooden tubs held 9 cwt of coal.

A group of men at lunch in Frog Lane Colliery c.1905; all are using the peg and ball oil lamps.

The compound haulage engine in the high seam. It worked over a distance of 6,000ft. Horses were only used near the pit bottom.

Two colliers undercutting the Hard seam. The coal was then brought down by wedging, explosives were not used on the faces.

Glossary

A.B Machine – A Longwall under cutting machine made by Anderson Boyes, which undercuts the coal seam.

Air-Course – An underground roadway kept open for the purpose of conducting fresh air into various parts of the mine.

Banksman – A workman at the top of the pit removes the full tubs from the cage and puts empty tubs into the cage.

Barren ground – An area with no coal.

Bell-Mould – A stone which is the stem of the plant Sigillaria, which has a larger diameter downwards. They have very smooth surface and they are apt to drop out suddenly.

Billet – A short wood prop which prevented the roof falling.

Boards – A certain type of stone which lies above the coal seam.

Branch – A stone-drift tunnel made in solid rock.

Brattice – A partition in the shaft, normally separated the pumping section from the coaling section.

Broken or Gob – The section of the mine where the coal has worked out. A Bristol and South Wales term used in South Wales until the coalfield closed in the 1980s.

Cart – A Bristol term for a small sledge which held about 1 1/2 cwt of coal. The hudge was also known as a cart.

Cementation – A system of drilling bore holes in to poor ground, and injecting cement into the boreholes.

Clift Roof – An Argillaceous shale.

Clod A Somerset term for a softish stone, which lies over the seam, and invariably falls after the coal is taken.

Cogs – A timber structure normally used to support the roof.

Colliery take – The area which had been or was intended to be worked from a particular colliery.

Crop Workings – The point where the coal seam shows on the surface.

Dipple – A heading driven towards the dip or deep.

Down dip – The area where the seams dips down away from the surface.

Drift – A sloping tunnel driven into the coal seams. Used where a shaft is not needed.

Engine Pit – A shaft on which the pumping engine is placed.

Gug – A self acting incline, sometimes had a hand operated windlass.

Hatchings – A Bristol term meaning a topple or top-hole, an underground road driven to the rise, in which coals are brought down to trams or horses.

Haystack Boiler – The original boiler which was the shape of the original round haystack, mainly used on Newcomen Engines.

Hudge – A large iron barrel for raising coal, which held from 10 cwt to 20 cwt.

Kibble – A later iron barrel which replaced the hudge, used until recently for shaft work.

Pan Floor – The fire clay lying under the coal seams worked at Radstock and the neighbourhood.

Pillar of coal – Unworked area of coal normally left under a church or large house.

Put – A small sledge for hauling coal from the face.

Running stage – A device used for pushing the hudge to where it was to be tipped. Was also a cover which pushed over the shaft top once the hudge had been raised.

Slant – Dipping tunnel driven into the coal measures, to work and remove output of mine.

Stall – Part of the Pillar and stall system where one workman and assistant would cut out a workings place at right angles to the heading.

Stone Drift – See branch.

Trunk road or Gate Road – Main underground roadway often used as air intake and carried conveyors or underground tram way.

Undercutting machine – A machine which cuts a slot in the lower section of the coal seam, allowing the collier to bring down the mass of coal.

Panels – An area of worked coal usually rectangular served by roads on either side of the panel known as gate roads.

Plough – A post war device for winning the coal, which was dragged along the coal face shearing off the coal.

Putt – A shallow wooden box used for hauling the coal from the face. Held from 3 to 6 cwt of coal, when empty it weighs 60 pounds.

Shides – The pumps in the shaft.

Tip – A small pit sunk from one seam to another, was also known as a cut in Bristol.

Trepanner – A later form of coal cutting machine not used in Somerset.

OMS – Output per manshift.

Washout – An area of ground in which the coal for various reasons is not present.

Windbore – The section pipe or lowest section of the pump, which rests in the sump and draws in water through perforations in the pipe.

Vearers – (sometimes spelt veerers) A man who unloads the hudges with the assistance of a crane, later became the banksman.

Many of these terms used in the Bristol Coalfield were still in use in the South Wales Coalfield until the premature closure in the 1980s. The terminology in Bristol and South Wales was the same because many of the South Wales families originally came from Bristol, Kingswood and Somerset.

 # Bibliography

Anstie, John, *The Coal Fields of Gloucetershire and Somersetshire and their resources*, (1873)

Berg, Torsten & Peter, *R. R. Angerstein's Illustrated Travel Diary, 1753-1755*, (2001)

Braine, Arthur, *A History of Kingswood Forest*, 1891

Cossham, Handel, *Statistics of the coal trade*; two lectures by Handel Cossham delivered at the Bristol Mining School, 22 June and 17 August 1857

Parliamentary reports 1) children's employment 1841,2) coal commission report 1871

Rogers, K. H. *The Newcomen Engine in the West of England*, 1976

Vinter, Dorothy, *Some Coalfields in the neighbourhood of Bristol and Kingswood,* 1964

Watkins, George, *Stationary Steam Engines of Great Britain*, vol 6, 2002

Index

A
Accidents 44, 47, 48, 67, 69, 73, 74, 100, 105, 113, 115, 116
Arthur, Charles 24
Ashton seams 34, 65
Avonside Level 73

B
Barrs Court Pit 17, 55
Belgium Pit 57, 59
Bell pits 5, 7, 8, 10, 11, 33, 55, 70, 79, 97, 118
Berkeley, Sir John 46
Bitton 5, 9, 23, 24, 25, 28, 110, 111, 114, 117, 126
Boulton & Watt engine 69
Brain, Aaron 47, 114, 115
Brandy Bottom Colliery 9, 102, 103, 104, 105
Brickyard Pit 33
Bristol Wallsend Collieries Limited 94
Brook Pits 74
Bryants Pit 100
Buff Pit 78
Buff Vein 8, 13, 37, 40, 78
'Bull' engine 52
Bull Hall Colliery 76

C
Cadbury Heath 7, 8, 78, 82, 86
California Colliery 8, 74, 78
Camerton Colliery 21, 74
Candles 21, 35, 45, 48
Capel ventilating fan 65
Capell fan 128
Captain Copley's iron-works 46
Carbide lamps 23
Chaffhouse show 105
Chester Park 24, 26, 46
Children in mines 33, 37, 44, 48, 56, 78, 81, 105, 115, 116, 128
Church Farm Colliery 97, 98
Church Fault 81, 86
Church Lease Pit 118
Clay Hill Colliery 49
Coalpit Heath 5, 7, 8, 9, 21, 24, 25, 89, 100, 109, 118, 127, 128, 130
Cock Road 7, 9, 14, 18, 21, 55
Cockshot Hill 70

Combe Brook 46, 47, 49
Conham Copper Works 25
Cooks Pit 100, 105
Cools Level 21
Coombe Brook Valley 21
Corn Horn Hill 25
Cossham, Handel 21, 47, 48, 61, 65, 66, 101, 102, 105
Cottles Engine Shaft 64
Crews Hole 8, 26, 40, 42
Crop Workings 5, 7, 97, 138
Crown Colliery 7, 81, 83, 84, 86, 87
Cuckoo Vein 79

D
Deep Pit 9, 31, 45, 47, 56, 58, 59, 60, 61, 64, 65, 66, 78, 97
Denglys Pit 45
Devils Vein 40
Donkham Pit 64
Doxall shaft 64
Doxall Vein 35, 45
Dudley Pits 100, 109
Duke of Beaufort 5, 19, 21, 24, 45, 47, 49, 50
Duke's Engine pit 45
Duncombe Pit 25

E
Easton Colliery 2, 4, 30, 32, 33, 34
Egg-ended boiler 65, 70, 97, 109, 117, 121, 128

F
Frampton Cotterell 7
Frog Lane Colliery 4, 22, 89, 109, 118, 127, 128, 134, 135, 136
Frog Lane colliery 109
Fussell, Abraham 74

G
Garden Pit 100
Garrotts Mead 70
Gee Moor 55
Gin 8, 9, 24, 35, 39, 40, 45, 56, 69, 76, 78, 86, 87, 97, 102, 104, 109, 115, 116, 118, 123, 130, 132
Glossary 138

Golden Valley 4, 7, 9, 24, 25, 28, 32, 44, 50, 94, 110, 111, 114, 115, 116, 117, 119, 121, 126
Goldney Pit 4, 25, 78, 82, 84, 86
Goose Green 8
Great & Little Dicot Pits 47
Great Cart Pit 100
Great Vein 25, 32, 33, 35, 45, 46, 48, 51, 52, 55, 56, 61, 63, 65, 67, 69, 70, 72, 73, 79, 89, 94, 97, 101, 105, 117
Great White Fault 21, 48, 52, 58
Gregory of Kingswood 33, 34, 35, 74
Grimsbury Pits 4, 79, 80, 81
Guibal fan 128

H
Hanham Colliery 7, 51, 52, 55
Harry Stoke Mine 89, 90, 91, 92
Hen Pit 50
Hole Lane Colliery 76, 77, 110, 117
Hollyguest Pit 17
Horses Underground 34
Hudds Pit 45

I
Iron Works 9, 55
Isaac Taylor 26

J
Jefferies Walters & Co. 102
Jones & Co. 79
Jones Pit 45

K
Kingswood 5, 7, 9, 21, 23, 24, 25, 26, 32, 33, 34, 35, 44, 45, 46, 47, 48, 51, 52, 55, 56, 58, 61, 64, 65, 66, 67, 69, 70, 72, 73, 74, 79, 87, 89, 101, 117, 139
Kingswood Great Vein 25, 32, 35, 45, 47, 48, 51, 56, 61, 65, 67, 69, 70, 72, 73, 89, 101, 117

L
Land Pit 97
Lapwater Pit 105
Lees Hill 70
Little Toad Vein 47, 69

Lodge Engine Pit 25, 26, 48
Lodge Hill 8, 19, 20, 21, 46, 47, 48, 69
Lodge Pit 19, 47, 67
Longwell Green 7, 12
Lord Radnor's Pit 102

M
Magpie Bottom 11, 21
Mangotsfield 7, 9, 46, 88, 94, 95, 97, 100
Mays Hill Pit 127, 128
Middle Engine Pit 24
Monks, John 47, 67
Mount Hill 55

N
National Union of Miners' 89
New Cheltenham Pit 10, 70
New Cock Pit 50
New Engine Pit 109, 118, 128
New Level 21, 25, 26, 46, 50
New Pit 67, 102, 103, 110, 111, 112, 113, 114, 115, 116, 118
New Rock Colliery 92
New Smith's seam 79
Newcomen Engines 7, 24, 25, 37, 67, 138
Nibley Colliery 26, 127
Nibley Pit 118
Norborne Berkely 24
Norton Hill Colliery 92

O
Old Brislington Pit 25
Old Doxall Pit 49
Old Engine Pit 25, 100, 118
Old Pit, Bitton 23, 24, 25, 28
Oldland Colliery Co. 74
Orchard Pit 109, 118, 128

P
Parkfield Collieries 65, 66
Parrot Seam 52, 76, 110, 116
Peg and Ball Lamps 21
Pendennis Pit 49

Pennywell Colliery 32, 33, 34
Pilemarsh Pits 26, 37
Plaited rope 9, 81
Players Levels 21
Pomphrey Farm 97
Potters Level 21, 70
Potters Wood and Jays Pits 70, 71, 73
Pucklechurch 26, 105
Pumping Engine 19, 23, 24, 25, 26, 28, 34, 37, 40, 44, 47, 48, 49, 52, 55, 65, 66, 74, 77, 78, 84, 87, 88, 99, 100, 115, 117, 121, 127, 134, 138

R
Rag Vein 8, 40, 73, 110
Ram Hill Engine Pit 118, 128
River Boyd 114
Roe Yate 21
Rotherham Colliery 4, 67
Royal Oak Pit 50
Rubens Pit 56, 59

S
Schiele Fan 65
Scotch Lamp 21
Serrage Engine 25
Serridge Engine 118
Shortwood Collieries 105, 106
Shortwood Engine Pit 100
Shortwood Lower Pit 105
Shoscombe Colliery 79
Sirocco fan 103
Siston Common Pits 81
Siston Engine 24
Siston Hill 8, 23
Smiths Pit 55
Soundwell 7, 8, 21, 23, 24, 26, 46, 48, 67, 68, 69
Soundwell Colliery 8, 23, 26, 67
South Liberty Colliery 23
South Pit 86, 103
Speedwell 9, 23, 25, 45, 47, 56, 58, 59, 62, 63, 64, 65, 66, 89, 92

Speedwell Colliery 9, 56, 62, 63, 66, 89, 92
Staple Hill railway tunnel 50
Star Pit 50
Starveall Pit 56
Stinking vein 49
Stones Pit 20, 47
Syston Common Engine 26

T
Tennis Court Pit 79
Thompson Pit 14, 15, 16, 55
Three Cocks Inn 39
Troopers Hill Fireclay mine 42
Troopers Hill Pit 40
Two Mile Hill 5, 7, 8, 21, 26, 44, 45, 52, 65, 76
Tylers Pit 45, 56

U
Upper Wood Pit 100, 105, 107

V
Ventilation Furnace 15, 55, 56, 80, 111, 112, 115, 116, 117, 122
Victoria Pit 50

W
Waddle Fan 34
Wages 33, 45, 46, 56, 78
Warmley 7, 9, 10, 16, 21, 23, 24, 25, 26, 78, 81, 86, 87
Warmley Colliery 7
Webbs Heath 7, 8, 81, 84, 85
Westerleigh 7, 118
Whitehall 7, 13, 32, 34, 35, 37
Whites Hill Pit 45
Winding engine 24, 33, 34, 35, 37, 38, 52, 54, 55, 56, 64, 65, 69, 74, 78, 86, 87, 95, 96, 97, 103, 105, 109, 111, 116, 117, 119, 127, 128, 130, 134
Wood Pit 100, 105, 107